CREATIVE
MATHEMATICS

African Institute of Mathematics Library Series

The African Institute of Mathematical Sciences (AIMS), founded in 2003 in Muizenberg, South Africa, provides a one-year postgraduate course in mathematical sciences for students throughout the continent of Africa. The **AIMS LIBRARY SERIES** is a series of short innovative texts, suitable for self-study, on the mathematical sciences and their applications in the broadest sense.

Editorial Board

AIMS Library Series

CREATIVE MATHEMATICS

A GATEWAY TO RESEARCH

ALAN F. BEARDON
University of Cambridge

CAMBRIDGE UNIVERSITY PRESS
Cambridge, New York, Melbourne, Madrid, Cape Town, Singapore,
São Paulo, Delhi, Dubai, Tokyo

Cambridge University Press
The Water Club, Beach Road, Granger Bay, Cape Town 8005, South Africa

Published in the United States of America by Cambridge University Press, New York

www.cambridge.org
Information on this title: www.cambridge.org/9780521130592

© Cambridge University Press 2009

First published 2009

Printed in the Republic of South Africa by Paarl Print

ISBN 978-0-521-13059-2 Paperback

Contents

Contents

Preface

This text is written to help readers experience the joy of creating their own mathematics. At its simplest level, the first half of the text can be read by students at school; the second half contains, or at least suggests, material at graduate level and beyond. The aim is to provide guidance (rather than instructions) for readers who are expected to be actively involved in attempting to solve the wide variety of problems in the text.

Part I consists of three essays: on solving problems, on writing mathematics, and on giving a presentation. These express the author's personal views, developed over many years, and it is hoped that the reader may find something new and useful in these.

The rest of the book contains a variety of problems, hints, generalisations and (sometimes partial) solutions. In more detail, this part, and indeed each problem, is organised as follows:

In Part II Some elementary problems are introduced, sometimes with suggestions for investigations.

In Part III The likely results of the investigations are discussed, and the problem is solved (usually in an elementary way). Possible generalisations of the problem are suggested, and directions for further investigations are given.

In Part IV Solutions of some of the generalisations are given, and yet more problems are raised. At this point, there may still be a substantial investigation ahead.

This is not a text about solving conventional problems. As the title suggests, it is about solving problems in the sense of doing research.

Perhaps the main difference between these two views is that in research, as in this text, obtaining a *prescribed* answer is not the primary aim; the aim is reach *some* conclusion and, if necessary, we should be prepared to change the problem in order to reach a conclusion. The objective is to use mathematics as best we can to understand a situation, and variations of it. The emphasis is on understanding the applications of mathematics, and not on finding the answer.

The problems in this book were developed over several years in a course on *Problem Solving* given by the author at the *African Institute for Mathematical Sciences* at Muizenberg, South Africa. The Institute was first opened in September, 2003, and it takes graduate students from all over Africa. In the course, the students are confronted with the problems, often stated in a rather vague way, and the discussion develops through group work, computer experiments, and class discussions. The students are expected to be creative, to formulate their own definitions where necessary, and to write computer programs to obtain numerical evidence relating to the problems.

The author believes that the ideal problem should have the following features.

- The problem can be given in simple, non-technical terms.
- There must be an easy route into an initial discussion of the problem, preferably with an opportunity to use a computer to obtain further evidence for or against a conjecture.
- The solution must contain some interesting mathematics; the purpose of solving problems is to motivate the reader to learn more mathematics. Problems that are solved by an ad hoc method are of little interest as they do not contribute to this end.
- The problem must lead to significant generalisations, so that the reader will see how an elementary investigation can, and should, lead to further discussion and conjectures.
- The generalisations should be either extremely difficult to solve, or possibly open questions. The reader must learn that most problems are 'open-ended', and only rarely can we say that we have 'finished' a problem.

Part I
Style and Presentation

1

Solving Problems

What do we mean by 'problem solving', and what benefits should we expect from solving problems? Of course, by solving a given problem we obtain an answer, from which we may be able to draw a conclusion. There are practical problems which arise in industry where it is important, perhaps from financial considerations, to find an answer, and there is only a secondary interest in how the answer has been obtained. In such cases, the answer is the main objective. While I hope that the benefits that accrue from studying this text will improve the reader's skills in solving such problems, we shall focus here on the *process* of solving problems, rather than on the answers. In fact, the answers will be of little interest to us, except in that they illustrate a method, or suggest further investigations.

The primary object of our study is the problem itself, and its main roles are to show us how mathematics can be applied in a variety of ways, to provide motivation for us to learn more mathematics, and to see and experience how simple cases lead to a greater understanding, and hence to further problems, generalisations, and so on. Solving problems in this sense is like a journey of exploration; we must constantly pay attention to the local details, but all the time be aware of how these details fit into a much larger, unknown, picture. It has been said that we do not understand a piece of mathematics unless we can generalise it, and a generalisation usually calls for different ideas. Thus we should see our attempt to solve a particular problem as a continuously evolving account of a wider problem.

Many educationalists favour this experimental approach, believing that one can only learn mathematics by doing mathematics oneself.

This has much to recommend it but, by itself, it cannot be enough. There is no doubt that to succeed in mathematics one needs a vast supply of mathematical knowledge, and one cannot be expected to provide this entirely by one's own efforts. At some stage we must learn from others (Newton's phrase 'standing on the shoulders of giants' springs to mind here), so what is the best way to learn more mathematics? This text is an attempt to show how problems can motivate this learning.

How should we proceed when faced with a non-standard problem? In general, we will not be asked to show that a certain formula is true; we will simply be asked to "Investigate the following situation: . . . ". Suppose that we have never seen any problem like this before; how do we begin? First, if we are to use mathematics, we need to formulate the problem in mathematical terms and sometimes this can be quite difficult. It is, however, an essential step and it is often the key to a solution. Sometimes the problem will not be properly defined; for example, if we need to describe a pattern of coloured discs, are we to regard two patterns as being the same if one is a rotation of the other? What if one is a reflection of the other? There is a decision to be taken here, and it must be emphasised that the responsibility of taking such decisions *lies with the problem-solver*. In truth, each decision will represent a different problem. Thus, when carefully formulated, the original problem is often transformed into several different problems.

Next, we should try to specialise the problem, or try simple cases; indeed, do anything that gives us more insight into the situation, and which enables us to 'tune in' to the problem. It is reasonable to suppose that the simpler cases will be easier to solve than the general problem, but there is a danger here for it is often possible to solve a simpler case by a method which will not be applicable in the more general case. As long as we are aware of this, no harm (but some good) will have been done. The results of simple cases may lead us to conjectures, and these should be checked wherever possible by use of a computer, geometry, graphics, sketches, rough calculations, and so on. If a conjecture turns out to be false, we should not be disappointed; we must find the error in our thinking and then more progress will have been made. Sometimes, the tools (matrices, groups, calculus, . . .) needed to solve a problem will be clear, but they may need to be refined. For example, we may suspect that we need something about numerical congruences, but that our knowledge is not enough. In this case an hour or so browsing though appropriate texts in the library might be helpful. If, at any stage, we realise that our problem, or proposed solution, has anything to do with some other part of mathematics, we should take time off to study this. Solving problems in the sense described in this text takes time, and the reader should not expect to solve the problem in one sitting. Experience shows that our thoughts about a problem will evolve with time, so we must give the problem the time it needs and deserves.

Suppose that at some stage we seem to have reached a dead end, and all our efforts have failed. Maybe this is because we are looking at the problem in the wrong way. One of the most attractive and powerful features of mathematics is the way in which the same thing can be said in many seemingly different, but actually equivalent, languages. It should not surprise the reader to learn that some of these languages are more easily adapted to solve a given problem than others, so perhaps the real issue is to find which of the languages is the best one to use. For instance, a problem involving integers might best be expressed in terms of congruences, or in terms of prime numbers, or in terms of binary numbers, or in terms of a group in modular arithmetic. Finding out which (if any) of these is the best to use is part of the skill in problem solving, so we should constantly think about the possibility that another language might be better suited to the problem under consideration. Very often, the process of generalisation requires us to change from a description in one language to another.

It is important that from time to time we write a carefully crafted account of what we have achieved, what we hope to achieve, and what we have tried, but failed, to do. This very exercise of putting our thoughts on paper in a coherent way will (provided that it is treated seriously) often give a further stimulus, or provide new ideas, for it requires us to think, and organise our thoughts, about our problem. We should take every opportunity to discuss the problem, and our ideas for a solution, with friends, colleagues and others. Ultimately, mathematics is about *sharing* ideas, for only in this way can we, as a group, extend our collective knowledge of mathematics in an efficient way.

Let us now suppose that we have solved the problem; have we finished? Surely not! First, have we really understood our solution, or do we have some irrelevant, or redundant, mathematics in it? If so, then we should rewrite the solution without it. Eventually, when we have removed this unneccesary material, we can concentrate on the heart of our solution. What have we used, and what have we assumed? Does this method solve a more general problem? If we change some of the parameters in the problem do we still have a solution? If so, we should acknowledge this; if not, then we have another problem which we should also try to solve. The reader should realise that once a problem is solved, there still remains the task of *reviewing, polishing, and rewriting the solution until it finally appears as an elegant piece of carefully argued mathematics*. One should not include one's first (probably wayward, sometimes incorrect, and often irrelevant) attempts in the 'final' write-up of the problem!

Finally, *a problem is never finished*, and it is for the reader to decide at which point it will be more useful to move to another problem, or more fun to do something else, or just relax.

2

Writing Mathematics

There are many circumstances in which we may be asked to write some mathematics or, more generally, a scientific report. It is not easy to write a good report, and it takes a lot of time to learn, but it is a skill that most people can learn. You may find that writing a report is more difficult than giving a lecture. When we give a lecture we can assess how the audience is reacting, and we can say more, or less, in response to this. We cannot do this when writing a report. Here we offer some general suggestions and hints on how to write a good report. We do not specify the nature of the report, and, for obvious reasons, we direct much of our attention towards the writing of mathematics (which we continue to refer to as a 'report'). This is *not* a set of golden rules to follow; rather, it is a list of points to think about, and consider, before, and during, the writing of the report. It should help us decide what to include and what to omit.

Before we consider the typical structure of a report, you should consider the following important questions.

- *Why are you writing a report?*
- *Who will read the report, and what are they looking for?*

In order to write a good report we must be clear about the purpose of our report: *what do we want to achieve, and what is our message?* Keep the purpose of the report in mind at all times; we should have an identifiable reason for including each item in the report.

We should *think carefully about the level of knowledge of our intended readers, and what they want to achieve by reading the report,* and write accordingly. It is very easy to forget who we are writing for, and at each stage of the writing we should ask what will the reader think of this? Also, be consistent; for example, do not solve quadratic equations while discussing an advanced topic in topology! At any given level some things are obvious and do not need to be said, while other things may need considerable explanation; in each case consider which of these is applicable.

Let us now consider the structure of a typical report. Most reports will have something like the following sections:

1. Title
2. Abstract
3. Contents
4. Introduction
5. The main body of the report
6. Conclusion
7. Acknowledgments
8. References

The first four sections should be designed to lead the reader gently into the main body of your report. We should include a statement of what we want to achieve, and a summary of our conclusions. These sections should give the readers a sense of direction; readers do not want to go on a journey with no idea of where they are going. A brief description of these sections now follows.

Title This should be informative (not, for example 'On a theorem in analysis').

Abstract The sole purpose of the abstract is to enable potential readers to decide whether or not they want to read the report. It should be entirely self-contained, and it should not be necessary to consult other works to understand the abstract. Usually, it will contain a brief non–technical description (without symbols) of the contents of the report.

Contents This tells the reader how the report is organised; it is a road map of the report. Choose the titles of each chapter/section carefully; they should give the reader a clear indication of what each chapter/section contains.

Introduction This should provide an easy entry point for the reader. The introduction usually includes some background material (sometimes the motivation for the report), a statement about what we are assuming that the reader already knows, and what we will achieve in the report. It is often an elaboration of the abstract, but these two sections have a different purpose. The abstract enables the reader to decide whether or not they want to read further; once they have decided to continue reading, the introduction is taken as an agreed starting point for the rest of the report.

The main body of the report This section should contain the bulk of the report and, broadly speaking, should follow the lines suggested in the Introduction. This part of the report should contain a discussion of the work that the report depends on, detailed arguments and analysis, and so on, leading to the conclusions. It is a good idea to insert figures last, because creating and changing figures can take a lot of time.

Conclusion This is a summary of what we have achieved, and we should state whether we have achieved the objectives as set out earlier in the report. If appropriate, some open problems or suggestions for further work can be included here.

Acknowledgments Here we acknowledge any assistance or joint work in the report.

References The references are numbered, and are usually given in alphabetical order (in mathematical writings), although there are other accepted conventions (for example, in the order in which they appear in the report). References in the body of the report are usually indicated by, for example, [2]. We are entitled to refer to, and use, results that are already in the literature but we must refer to them explicitly and list the reference in this section.

We end this section with some points to consider when writing the report. First, the structure of the report should be coherent, and it should make it easy for the reader to follow the arguments. Before we start writing we should sketch a brief outline of the structure. We should think about what to include in each section, and we should ensure that we have a reason for including each item. Also, we should think about the length of each section. It is a good idea to write a rough contents page and allocate page numbers. It doesn't matter at all if this allocation changes as we write, but it will act as a guide while we are writing the report. Moreover, if there is a length restriction, this will help us conform to it.

Pay attention to details of the structure of the report. A new section indicates a strong division of the material; a new paragraph indicates a much weaker division. Within a section, paragraphs should be loosely related. As we read/write the report we should ask ourselves whether there are too many different ideas in a paragraph. Generally, a sentence should express only one idea. Be aware of the number of ideas in a sentence or paragraph. If we introduce too many ideas at once, the reader will find it difficult to understand what is written. Also, do not give the reader information before they need it for they will have probably forgotten it by the time we refer to it.

In some ways writing a report can be more difficult than giving a lecture. When giving a lecture we can immediately assess whether or not the audience understands what we are saying, and we can then choose to say more or less as appropriate; indeed, we can even repeat ourself (several times if necessary), which we should not do in a report. Throughout the report we must be careful to say clearly and exactly what we mean so that the readers do not require further, or a repeated, explanation. We should expect to rewrite the report many times before we get it right.

There are several ways to write the main body of the report. Some people begin writing the report with the main body because, in their view, it is easier to write the Introduction and Abstract after this. Other people start with the first line of the report and then, after a while, go back to the beginning and re-write all that has been written so far (repeating this process many times). Yet another way, if we think of the report as a human body, is to build the skeleton first and then add the flesh to the report. If we feel that a section is getting too long and complicated it is a good idea to inform the reader where we are in the grand scheme of things. It is the author's responsibility to tell the reader where they are at present, and where they are going next. We should inform the readers often enough to keep them comfortable (and we will need to rewrite the report several times to get this right).

It is important that the report is grammatically correct; punctuation and spelling are just as important in mathematics as in literature (and perhaps more so), and the report will be much easier to read and understand if it is well written. Mathematics is about precision, and we cannot be precise if the punctuation is poor; in particular, formulae, including displayed formulae, are part of a sentence and need correct punctuation. Mathematicians are good at converting ideas into symbols, but this does tend to concentrate many ideas into a short space. We must be aware of this when writing the report, and not overload the reader with ideas. Often a few words (rather than a few symbols) will explain an idea more clearly.

Be precise in mathematical statements, and be careful at every step; we cannot be half right in mathematics. Take care to say *exactly* what is meant, and avoid vague statements. We do not have to include every detail, but there must be no ambiguity in our statements. When we make an assertion it should be absolutely clear why we are entitled to make it. For example, is the assertion obvious, or trivial, or a consequence of a (possibly) famous theorem, or a consequence of a few logical steps from something we have already stated? The reader should not be left in any doubt about the status of an assertion. At each stage try to give the reader enough information to prevent them breaking their train of thought; this can often be done by reminding them of a previous definition or theorem.

As you write, remember that everything you do should have a purpose. Keep asking yourself, why am I including this? Why am I structuring this section in this way? Is what I have written clear? You should constantly re-read and re-write your report. Get other opinions on it; give it to other people to read. Finally, after you have finished the report leave it for a few days, do something else, and then re-read it. Is it still clear; does it flow well; does it say what you want it to say? If so, then you are finished.

3

Giving a Presentation

Giving a good presentation is difficult. If a presentation does not go well don't be discouraged, but make sure you learn from the experience. All of us have given a good presentation, and a bad presentation. Here are some guidelines which might help you to deliver a good presentation. First you have to prepare for the presentation; then you have to present it to the audience.

The preparation

First think about who your audience will be; this is very important and will govern what you say in your talk. If you are giving a mathematical lecture it is important to know the age and stage of development of the audience; you would give a very different talk to children than to graduate students!

Decide upon your message: what is the purpose of your talk? What do you want the audience to know and understand? Plan how to get your main points across effectively. It is best to begin with something simple or familiar to everyone and then build up to a difficult point later. Choose the main points of your talk carefully, and consider how much time you have to explain these points, and what you can reasonably expect your audience to understand. A good way to do this is to make a list of what you *might* include, and then eliminate some of the items from the list until you have a reasonable amount of material left. Do not give too much, or too little, information: if you give too little the audience's thoughts will wander, while if you give too much the audience will stop concentrating. It is also very important to decide *what not to tell your audience* – some things are too complicated to include in a talk, and you will not have time to explain everything you know about the subject.

It is best to prepare more than you think you will need; then you will not run out of things to say if the earlier material took less time than you thought it would. In addition, if the audience seems confused you will be able to switch to a different topic; if they seem to be finding the material difficult, you can switch to an easier topic.

The presentation

Perhaps the most important point is to *time your presentation carefully*. Do not rush at the beginning (even if you are nervous), for if you lose the audience at the start they will not concentrate fully for the rest of the talk. Also, do not spend too long explaining one point, even if the audience appears not to understand it. If you are using prepared slides, or electronic equipment, beware: *there is always a tendency to present the material too quickly*. One advantage of writing during the lecture is that it slows the lecturer down to a pace that is more suited to the audience. Above all, *keep the time spectrum in your mind, and remember to pace yourself*. Throughout the lecture you should be thinking about what you said five minutes ago, what you are saying now, and what you want to say in five minutes time. This is not an easy skill to master!

If you are giving your presentation in a language other than your own, or to people who would not normally speak your own language, you will need to prepare for your presentation even more carefully. Keep your language as simple as possible, and deliver the talk at a slower pace than you normally would.

Try to relax and be natural; this is easier to do if you can speak without notes. If your manner is friendly and relaxed the audience is more likely to feel comfortable, and to concentrate on what you are saying. The first sentence is often the most difficult (especially as you may feel nervous at the start), so it is worthwhile to memorise a simple opening sentence or two. Do not be too formal, for this will make the audience feel uncomfortable. *Make eye contact with individuals in the audience* frequently; this will allow you to assess whether or not the audience understands what you are saying. If people look confused you can slow your pace, or omit difficult material; if people look bored you can speed up.

Make sure that you pause after a difficult point to give the audience time to absorb what you have said. Remember that although you may have rehearsed this many times, it is probably new to them. However, *you must time the length of the pause carefully*; if it is too short it will not serve its purpose; if it is too long, the audience will start to think of other things. Do not think about other matters during the pause; it will disconnect you from the audience (or even your talk). Instead, spend the time watching and assessing the audience; the pause should be of benefit to you as well as the audience.

You will have a much better chance of holding the audience's attention if you move around during your talk. Also, vary the pace, tone and volume of your voice. Move into a quieter mode just before an important point that you want the audience to remember; then you can raise your voice, or wave your hands, in order to emphasise the point. Don't be afraid to repeat an important point – several times if necessary. Think of yourself as an actor; a good actor engages the audience.

It is clear that the way you present your conclusion is most important. Make sure that you leave enough time to summarise your main ideas and leave the audience with a clear but simple message to take away. Do not introduce new ideas at this stage except possibly to suggest further avenues for investigation.

Finally, it is hardly necessary to say that you can only follow these suggestions if you are totally familiar with the content of the talk. For most of us these suggestions demand our full attention, and there is usually very little mental energy to spare to think about the content of the talk while giving it.

Part II
The Problems

The problems we examine can be roughly classified as follows. Problems A, C, E, and perhaps J, require linear algebra, matrices and so on. The essence of these problems is to realise that the essential features are of a linear character, though not necessarily over the field of real numbers. Elementary cases (where the parameter is small) can be solved by simple matrix computations, while a discussion for larger values of the parameter introduces the idea, and the problems, of finding large matrices on the computer.

Problems B, G and I, which at first sight look elementary, require number theory and, in particular, will develop a good understanding of binary numbers and modular arithmetic.

Problems D, F and H are of a probabilistic nature. Two of these show the power of generating functions, and these go beyond what is normally taught at this level.

Finally, Problem K requires a little real analysis of a practical sort, combined with a critical look at what appears on our computer screen.

4

A First Look at the Problems

Problem A: Circles and triangles

Before you read further, find a necessary and sufficient condition for three positive numbers L_1, L_2 and L_3 to be the lengths of the sides of some triangle. Now justify your claim.

- *Given a triangle in the plane, is it always possible to construct three circles, with their centres at the vertices of the triangle, such that each circle is tangential to each of the other circles?*

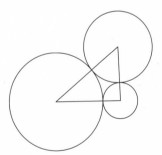

Fig. 1. Three tangential circles

Figure 1 illustrates one such set of circles. Try some special cases in which you know the lengths of the sides of a triangle: for example, right-angled triangles with sides of lengths 3, 4, 5, or 5, 12, 13. What other Pythagorean triples are there? Now try to solve the problem for a general triangle with sides of lengths L_1, L_2 and L_3. You should consider the case where all of the circles are exterior to each other, and the case where some of the circles lie inside another. If you give an algebraic solution then you must show that the radii are positive.

Problem B: Towers of positive integers

Consider a positive integer, say 4937. The numbers 4, 9, 3 and 7 are the *digits* of 4937, and 7 is the *last digit* of 4937. Now let x be any positive integer. The numbers x, x^2, x^3, ... are the *powers* of x, and the numbers

$$x, x^x, x^{x^x}, \ldots$$

are called the *towers* of x.

* *Given that $x \in \{1, 2, \ldots, 9\}$, what is the sequence of last digits of the towers of x?*

Normally, in a numerical problem, we consider each case, make some calculations, look for patterns, and obtain further evidence from a computer. We then make some conjectures and try to prove them. However, in this case, a few rough calculations show that the towers of x grow very rapidly.

* *How many digits do 10^{10} and $10^{10^{10}}$ have?*

The towers of x grow too rapidly to compute, so you may have to think a little more about this problem in order to make any progress. The last digit d of x satisfies $x \equiv d \pmod{10}$, so perhaps we can use congruences to reduce the numbers to a manageable size? This might mean that you need to learn more about congruences in order to solve this problem.

Problem C: Coloured discs

We are given n discs, where $n \geq 2$, and each disc is either red or blue. We form the *first pattern* by placing the discs at equally spaced intervals around a circle. We then obtain the second pattern by (i) placing a red disc between any two adjacent discs of the same colour in the first pattern, (ii) placing a blue disc between any two adjacent discs of different colours in the first pattern, and then (iii) removing the discs in the first pattern. The discs that remain form the *second pattern*. This process is repeated to obtain the third pattern, the fourth pattern, and so on. A sequence of first, second and third patterns is illustrated in Figure 2 in the case $n = 4$. Note that there are two variables here, namely the number of discs, and the different possibilities for the first pattern.

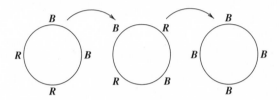

Fig. 2. First, second and third patterns

- *Can you analyse the cases $n = 2, 3, 4, 5$? Can you describe how the system will evolve?*

Often, the major difficulty in a problem lies in understanding how to formulate the problem in a mathematical way, and one of the main difficulties here is to invent a way of recording the changing patterns. Clearly we should regard two patterns as being the same if one can be converted to the other by a rotation, but should we regard them as the same if one can be converted to the other by a reflection?

Problem D: Throwing dice

The standard die (the singular is *die*; the plural is *dice*) is a cube with 1, 2, 3, 4, 5 and 6 black spots on its six faces (see Figure 3). For our purposes it is better to consider a cube with the numbers 1, 2, 3, 4, 5, 6 on its faces. In a certain game we throw two dice and add the scores on the two top faces. The only possible scores are 2, 3, ..., 12, and these occur with frequencies 1, 2, 3, 4, 5, 6, 5, 4, 3, 2, 1, respectively (prove this).

- *Can you create a new pair of dice with positive integers on their faces, which are different from the standard pair of dice, but which have the same sums with the same frequencies as a pair of standard dice? (You may use an integer on more than one face.)*

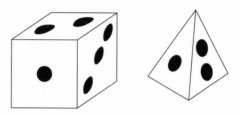

Fig. 3. Cubic and tetrahedral dice

We can consider a slightly simpler problem with tetrahedral dice. A *tetrahedral die* is a regular tetrahedron with its four faces labelled $1, 2, 3, 4$ (see Figure 3). When we throw a tetrahedral die the score obtained is the number on the face that is in contact with the table (and which we cannot see). If we throw two tetrahedral dice and then add their scores, the possible scores are $2, 3, \ldots, 8$ with frequencies $1, 2, 3, 4, 3, 2, 1$, respectively. For example, a score of 6 can only occur from the pairs $(2, 4)$, $(3, 3)$ and $(4, 2)$.

- *Can you create a non–standard pair of tetrahedral dice, with positive integers on their faces, that have the same sums, with the same frequencies, as a pair of standard tetrahedral dice?*

Problem E: Switching lights

Consider n lights, each of which can be ON or OFF. Each light has a switch, and if the switch on one light is operated then *all of the other* lights change from OFF to ON, or from ON to OFF, but *the light attached to the switch does not change*. For example, if there are three lights in the state (ON, OFF, OFF), and we use the middle switch, then the state of the lights becomes (OFF, OFF, ON).

- *If all of the lights are on, can we operate a sequence of switches to turn them all off? We may use each switch as many times as we wish, and in any order. Does the effect of operating several switches depend on the order in which they are used?*

- *Write a computer program to experiment with this problem. The input should be a given state of the lights, and which switch to use; the ouput should be the new state of the lights.*

Problem F: Triominoes

A *monomino* is a thin square piece of wood with each side of length one; a *domino* is a thin rectangular piece of wood with sides of lengths one and two; a *triomino* is a thin rectangular piece of wood with sides of lengths one and three. A $p \times q$ board is a rectangular board with sides of lengths p and q that is divided into pq squares in the obvious way.

- *Is it true that the only way to cover a 5×5 board with one monomino and eight triominoes is to place the monomino at the centre of the board?*

- *For which positive integers p is it possible to cover a $p \times p$ square board, without overlapping, by several triominoes and exactly one monomino?*

Problem G: Weighing chemicals

I have some scales, and weights of $1, 2, 3, \ldots, N$ grams. To weigh X grams of a chemical, where $X \in \{1, 2, \ldots, N\}$, I put the weight of X grams on one side of the scales, and balance the scales by putting the chemical on the other side. Now I do not need all of the N weights; for example, I can discard the weight of 3 grams (and use the weights of 1 and 2 grams instead), and I can also discard the weight of 6 grams (and use the weights of 2 and 4 grams instead). It is clear that I can discard even more of the weights and still be able to weigh any amount from 1 to N grams.

- *What is the largest number of weights that I can discard, but still be able to weigh any integer number of grams from 1 to N, and which weights should I keep?*

Sometimes it is useful to reformulate a problem in a different way. Instead of starting with a collection of weights and discarding some, I can start with no weights at all and then see which weights I need. This is an equivalent problem, but it seems much easier to get started on. Try some cases in which N is small. When you have what you believe to be the correct answer, try to prove that the number of weights you have used is minimal for that N.

Problem H: Average lengths

For *non–negative* integers a and b, the set

$$S(a, b) = \{(x, y) : a \leq x \leq a + 1, b \leq y \leq b + 1\}$$

is the square (including its boundary) in the plane whose sides have length 1, and whose bottom left–hand corner has co-ordinates (a, b). Let k be a positive number (not necessarily an integer), let $L(k)$ be the segment with end-points $(k, 0)$ and $(0, k)$, and let $N(k)$ be the number of the squares $S(a, b)$ that meet $L(k)$. Also, for each a, b and k, let $\ell(k; a, b)$ be the length of the segment $L(k) \cap S(a, b)$. The segment with length $\ell(\frac{5}{2}; 1, 1)$ is illustrated in Figure 4.

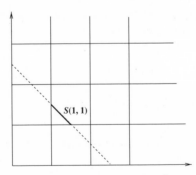

Fig. 4. The segment with length $\ell(\frac{5}{2}; 1, 1)$

- *Write a computer program that calculates $N(k)$ from an input k. Then find a formula for $N(k)$, and check several cases to provide evidence that the two methods give the same answer.*

- *What is the length $\ell(k; a, b)$? What is the average $A(k)$ of all lengths $\ell(k; a, b)$ taken over those squares $S(a, b)$ that meet $L(k)$? Does $A(k)$ tend to a limit as $k \to \infty$?*

Problem I: Diophantine triples

Diophantus of Alexandria (who lived around 250 AD, and who is often called the 'father of arithmetic') introduced the theory of *Diophantine equations*. These are equations with integer coefficients in which we seek only integer solutions. Diophantus noticed that the numbers

$$a_1 = \frac{1}{16}, \quad a_2 = \frac{33}{16}, \quad a_3 = \frac{68}{16}, \quad a_4 = \frac{105}{16}$$

have the property that if $i \neq j$ then $a_i a_j + 1$ is the square of a rational number (check this). We shall consider a similar situation for integers rather than rational numbers: a *Diophantine pair* is a pair of integers a and b such that $ab + 1$ is the square of an integer.

- *Can you find some Diophantine pairs or, better still, find a systematic way to list all Diophantine pairs? Does every positive integer occur in some Diophantine pair?*

A *Diophantine triple* is a triple of integers a, b, c such that $ab + 1$, $bc + 1$, $ca + 1$ are all squares of integers.

- *Can you find some Diophantine triples or, better still, infinitely many Diophantine triples?*

Do not be fooled into thinking this is a simple topic. Diophantus found a Diophantine 4–tuple (with rational numbers, not integers), and Fermat found a Diophantine 4–tuple of integers. However, we still do not know whether a Diophantine 5–tuple of integers exists or not!

Problem J: Midpoints

A triangle in the plane has vertices U, V and W, and A, B and C are the midpoints of the sides UV, VW and WU, respectively. If I give you the points A, B and C, but not the points U, V and W, can you find where U, V and W are? More generally, we ask the following question.

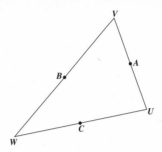

Fig. 5. The midpoints of a triangle

- *Given n points in the plane, is there an n–gon such that these points are the midpoints of the sides of the n–gon? If so, is the n–gon unique?*

Problem K: A five between zeros

If n is a large positive integer then

$$X_n = 10^n - \sqrt{10^{2n} - 1}$$

is the difference of two large numbers.

- *Is X_n large or small? Give a rough approximation of X_n.*

- *What can you say about the decimal expansion of X_n?*

Part III

Solutions and More Problems

5

Problem A: Solution

Suppose that $0 < L_1 \leq L_2 \leq L_3$. Then there is a triangle with sides of lengths L_1, L_2 and L_3 if and only if $L_3 \leq L_1 + L_2$. We leave the proof to the reader, but we remark that it is better to express this inequality in the form

$$2\max\{L_1, L_2, L_3\} \leq L_1 + L_2 + L_3,$$

as this does not require the L_j to be labelled in any particular way.

Many mathematical problems have two parts: (i) does a solution exist, and (ii) if it exists is it unique? Sometimes, if a solution is not unique we can characterise the set of all solutions in some way. The usual way to tackle the problem of existence is first *to assume that the solution does exist*. Then, working from this assumption, we gain some insight into some of the properties that the solution must have, and sometimes we can even find an explicit solution. After we have found what we believe to be a solution *it is essential to check that our answer is indeed a solution*. This is the only way to be sure that *our final answer does not depend on the earlier assumption* (which has served its purpose in giving a greater insight into the problem).

Let us now return to the problem of the circles and ask (i) does such a set of circles exist, and (ii) if it does, are the radii uniquely determined by the triangle?

First, we consider the case when *each circle lies outside the other two circles*. Let the lengths of the sides of the triangle be L_1, L_2 and L_3, and suppose that we draw circles of radii r_1, r_2 and r_3, with the circle of radius r_j centred at the vertex of the triangle opposite the side of length L_j (see Figure 6). The circles are externally tangent to each other if and

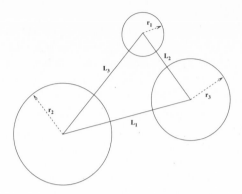

Fig. 6. Circles centred at the vertices

only if there is a solution to the simultaneous equations

$$r_1 + r_2 = L_3,$$
$$r_2 + r_3 = L_1,$$
$$r_3 + r_1 = L_2,$$

in which all of the r_j are positive. This system of equations has a unique solution given by

$$2r_1 = L_2 + L_3 - L_1,$$
$$2r_2 = L_3 + L_1 - L_2,$$
$$2r_3 = L_1 + L_2 - L_3.$$

The triangle inequality shows that each r_j is positive unless the triangle is 'flat' (that is, with collinear vertices), a case which we ignore for the moment. Note that a complete solution requires us to give the reason why the r_j are positive (even though it is trivial); in mathematics, *trivial statements are just as important as deeper statements since the omission of either type invalidates the logical structure of the argument.*

    The uniqueness (but not the positivity) of the solutions can be seen by noting that the system of equations above can be written as

$$\begin{pmatrix} 0 & 1 & 1 \\ 1 & 0 & 1 \\ 1 & 1 & 0 \end{pmatrix} \begin{pmatrix} r_1 \\ r_2 \\ r_3 \end{pmatrix} = \begin{pmatrix} L_1 \\ L_2 \\ L_3 \end{pmatrix},$$

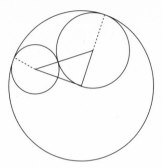

Fig. 7. Internally tangential circles

and that the determinant of the matrix of coefficients is non-zero. You should know how to compute this determinant by hand, and also by using a computer. We have now shown that *providing the triangle is not flat – i.e. not all its vertices lie on a straight line, there is a unique solution to the problem when all of the circles lie outside each other.* Now draw a triangle on paper, measure the lengths, solve the equations, and draw the circles to verify that this method really does work!

We must now consider the situation in which one of the circles contains the other two, and is still tangent to both of them (see Figure 7). With a suitable labelling of Figure 7 (which the reader should provide) the equations are now $r_1 - r_2 = L_3$, $r_2 + r_3 = L_1$ and $r_1 - r_3 = L_2$. This system of equations has the unique solution

$$r_1 = \tfrac{1}{2}(L_1 + L_2 + L_3),$$
$$r_2 = \tfrac{1}{2}(L_1 + L_2 - L_3),$$
$$r_3 = \tfrac{1}{2}(L_1 - L_2 + L_3),$$

and, providing the triangle is not 'flat', each r_j is positive. Thus our problem again has a unique solution.

We are still not finished. To complete our discussion we must consider other possible configurations of circles. For example, could all three circles be tangent to each other at the same point? You should now consider any additional cases (where you might allow one or more of the radii to be zero), and then prove that you have considered all possible cases.

- *We can ask a similar question if we replace the triangle by a general quadrilateral, or even an n–gon. What can we then say?*

Finally, formulate, and try to answer, the appropriate question in three dimensions: start with a tetrahedron in $\mathbb{R}^3$ (see Figure 8) instead of a triangle, and use spheres instead of circles.

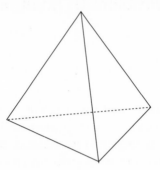

Fig. 8. A tetrahedron

- *Can you formulate, and solve, the analogous problem for a tetrahedron in 3 (or even n) dimensions?*

6

Problem B: Solution

First, you should have noticed that a tower of x is *not well defined*; for example, 2^{1^3} could mean $2^{(1^3)}$ or $(2^1)^3$, which are 2 and 8, respectively. This confusion arises because of the common, but careless, way of writing exponentiation *without an explicit symbol for the operation*. If we mimic a computer language and write $a \wedge b$ for a^b, then we have to decide whether $a \wedge b \wedge c$ means $a \wedge (b \wedge c)$ or $(a \wedge b) \wedge c$. There is an important lesson here: *we often have to create, or modify, a definition in order to solve a problem*. Here, as $(a \wedge b) \wedge c$ is already defined as a^{bc}, *we define*

$$a^{b^c} = a \wedge (b \wedge c) = a^{(b^c)}.$$

Now that we have properly defined repeated exponentiation, we can give a *formal definition of the towers of x by induction*: let $x_1 = x$, and, for $k = 1, 2, \ldots$, let $x_{k+1} = x^{x_k}$. *We shall use this notation x_k for the towers of x throughout the discussion*. For example,

$$x_4 = x^{(x^{(x^x)})}.$$

It is easy to see that the towers of x grow very quickly. As 10^n has $n + 1$ digits, it follows that $10^{10^{10}}$ has $10^{10} + 1$ (ten thousand million and one) digits. Also, $6^6 = 46656$, and if we put $6^{6^6} = 10^n$ and use logarithms, we see that 6^{6^6} has more than 36000 digits! It seems unlikely, then, that we can carry out numerical experiments on this problem. These remarks raise the following subsidiary problem.

- *Given a, b and c, estimate the number of digits of $a^{(b^c)}$.*

Next, we make two preliminary observations. First, when $x = 1$ all towers of x are also 1. This is trivial, but *a complete solution must contain a discussion of every case, including the trivial ones.* Second, since every tower of x is a power of x, *if x is even then every tower of x is even*, while *if x is odd then every tower of x is odd*.

The last digits of the sequence of *powers* of x may (or may not) be easier to find than the last digits of the towers of x. In any case, as this sequence contains all the information we want, we should consider this sequence too. Indeed, some of our results may hold for this more general sequence. *In general, it is advisable to consider related problems, especially if these seem to be easier than the original problem.*

We shall now use congruences to try to reduce the towers, or perhaps their exponents, to a manageable size. If $x = 5$ or $x = 6$ then $x^2 \equiv x$ (mod 10). Thus, by induction, $x^n \equiv x$ (mod 10) for all n. This shows that *if x is 5 or 6, then all powers (and hence all towers) of x end with the digit x.*

The next simplest case is $x = 9$. In this case, every even power of x ends in the digit 1, while every odd power of x ends in the digit 9. Thus the last digits of $9, 9^2, 9^3, \ldots$ are $9, 1, 9, 1, 9, \ldots$. As 9 is odd, every tower of 9 is an odd power of 9; so *every tower of 9 ends in the digit 9.*

The remaining cases are harder so, in the hope of generalising the argument given for $x = 9$, let us rewrite it in a more formal way. First, we noted that

$$9^n \equiv \begin{cases} 1 & \text{if } n \text{ is even;} \\ 9 & \text{if } n \text{ is odd;} \end{cases} \pmod{10}.$$

When $x = 9$, every tower of x is odd, so that $x_n \equiv 1$ (mod 2). This means that we can write $x_n = 2m + 1$, say, and then

$$x_{n+1} = x^{x_n} = x^{2m+1} \equiv 9 \pmod{10}.$$

Again, we see that *every tower of 9 ends in the digit 9.*

Next, suppose that $x = 3$ or $x = 7$. Then $x^4 \equiv 1$ (mod 10), so if $x_n = 4m + a_n$, where $a_n \in \{0, 1, 2, 3\}$, then

$$x_{n+1} = x^{4m+a_n} \equiv x^{a_n} \pmod{10},$$

and we can compute this. Thus we need to find a_n, which is $x_n \pmod 4$. As $x \equiv -1 \pmod 4$, and x_n is odd, we see that for $n \geq 1$,

$$x_{n+1} = x^{x_n} \equiv (-1)^{x_n} = -1 \equiv 3 \pmod 4.$$

This is also true for x_1, so that $a_n = 3$ for all $n \geq 1$. Thus

$$x_{n+1} \equiv x^3 \equiv \begin{cases} 7 & \pmod{10} \quad \text{if } x = 3; \\ 3 & \pmod{10} \quad \text{if } x = 7. \end{cases}$$

We conclude that *the sequence of towers of 3 ends in the digits 3, 7, 7, 7, 7, ...,* and *the sequence of towers of 7 ends in the digits 7, 3, 3, 3, 3, ...*

In the remaining cases, namely $x = 2, 4, 8$, x is even, so that if $n \geq 2$, then x_n is divisible by 4. Thus, for $n \geq 3$, $x_n = x^{x_{n-1}} = x^{4k} = (x^4)^k$, say. Now each of 2^4, 4^4 and 8^4 ends in the digit 6, so that $x^4 \equiv 6 \pmod{10}$. Thus, for these values of x,

$$x_n = x^{x_{n-1}} = x^{4k} = (x^4)^k \equiv 6^k \equiv 6 \pmod{10}.$$

Note that we have used the more general result that all *powers* of 6 end in the digit 6. Since this was not part of the original problem, *this shows the value of widening our view of the original problem.*

In conclusion, we have shown that for $n \geq 3$,

$$x_n \equiv \begin{cases} 1 & \pmod{10} \quad \text{if } x = 1, 9; \\ 6 & \pmod{10} \quad \text{if } x = 2, 4, 6, 8; \\ 7 & \pmod{10} \quad \text{if } x = 3; \\ 5 & \pmod{10} \quad \text{if } x = 5; \\ 3 & \pmod{10} \quad \text{if } x = 7. \end{cases}$$

The answers when n is 1 or 2 can be found by inspection.

New problems often arise during the discussion of a given problem, and you should pay attention to these as they often give us a better understanding of the original problem. For example, the original problem can be generalised as follows.

- *Let x be any positive integer. What are the last digits of the towers of x?*

Finally, we should ask whether there is any deeper number theory that lies behind these results. There probably is, and we should try to find out what it is.

7

Problem C: Solution

We can describe each pattern by a vector in which each component is R or B; for example, (R, R, B, B, B, B, R, B) means that $n = 8$ and, as we move around the circle in a clockwise direction, we see the colours red, red, blue, and so on, in this order. The starting point of this list is not well defined, so we agree that two patterns will be considered to be the same if one pattern can be rotated to the other. This means that we identify any vector v with those vectors that are obtained by cyclically permuting the components of v. Now let us consider the cases $n = 2, 3, 4, 5$.

The case $n = 2$ There are three possible patterns, namely $(R, R), (R, B)$ and (B, B), and the system evolves as follows:

$$(R, B) \to (B, B) \to (R, R) \to (R, R).$$

Regardless of the first pattern, after (at most) two steps we reach the pattern (R, R) and remain there.

The case $n = 3$ There are four possible patterns, namely (R, R, R), $(R, R, B), (R, B, B)$ and (B, B, B), and the system evolves as follows:

$$(B, B, B) \to (R, R, R) \to (R, R, R)$$
$$(R, R, B) \to (R, B, B) \to (R, B, B).$$

Here the pattern does not change after the first step, and it is then fixed at either (R, R, R) or (R, B, B). Notice that the final outcome depends on the first pattern, and that there are two distinct systems here.

The case $n = 4$ Here there are six possible patterns:

$$s_1 = (R, R, R, R), \quad s_2 = (R, R, R, B), \quad s_3 = (R, R, B, B),$$
$$s_4 = (R, B, R, B), \quad s_5 = (R, B, B, B), \quad s_6 = (B, B, B, B).$$

and you should check the following scheme:

Here the outcome does not depend on the first pattern.

The case $n = 5$ There are eight possible patterns:

$$s_1 = (R, R, R, R, R), \quad s_2 = (R, R, R, R, B),$$
$$s_3 = (R, R, R, B, B), \quad s_4 = (R, R, B, R, B),$$
$$s_5 = (B, B, B, R, R), \quad s_6 = (B, B, R, B, R),$$
$$s_7 = (B, B, B, B, R), \quad s_8 = (B, B, B, B, B),$$

and again there are two distinct systems, namely

$$s_8 \quad \rightarrow \quad s_1 \quad \rightarrow \quad s_1,$$

and

$$
\begin{array}{c}
s_6 \\
\downarrow \\
s_7 \\
\swarrow \qquad \nwarrow \\
s_2 \rightarrow s_3 \qquad \rightarrow \qquad s_4 \leftarrow s_5.
\end{array}
$$

Note that a new feature has emerged, namely that the pattern may end in a cycle; in this case it cycles between s_3, s_4 and s_7.

- *Can you explain immediately, without writing anything down, why we must always end up with some cyclic behaviour, regardless of the value of n, and of the starting pattern?*

Even if you have solved a problem in the sense that you have a complete understanding of what is happening, it is sometimes still difficult to express the results clearly and simply, *and in a mathematical framework.*

- *Invent a function that expresses the change of pattern in a mathematical way.*

   You should continue to experiment, but here are some general points for investigation.

- *The fixed states (the patterns which do not change) are of interest: what are they? Show that any fixed pattern must contain an even number of blue discs.*

- *Do you notice anything if n is of the form 2^m?*

- *How many different starting patterns are there?*

8

Problem D: Solution

First, we consider the simpler problem of two *tetrahedral* dice with positive integers on their faces. As the sum of the scores is at most 8, none of the integers on the faces can be greater than 7. Thus there are only a finite number of possibilities and we could use a computer to list all possibilities with the correct probability distribution. However, it might be more efficient (and is probably more interesting) to argue directly.

We label the first die with the integers a, b, c, d, and the second die with the integers A, B, C, D, where

$$1 \leq a \leq b \leq c \leq d, \quad 1 \leq A \leq B \leq C \leq D,$$

and we assume that this gives us a solution to the problem. As we can obtain a total of 2 in exactly one way, we must have $a = A = 1$, and $b \geq 2$ and $B \geq 2$. Also, as we can obtain a total of 8 in exactly one way, we must have $d + D = 8$, $c < d$ and $C < D$. Thus

$$a = 1 < b \leq c < d, \quad A = 1 < B \leq C < D = 8 - d.$$

Next, as we can obtain a total of 3 in exactly two ways, we must have exactly two of the labels equal to 2. Either each die has exactly one label of 2, or one die, say the first, has two labels of 2. Thus we now have these two possibilities:

$$\{1, 2, c, d\}, \quad \{1, 2, C, D\}, \quad 3 \leq c < d, \; 3 \leq C < D, \quad (1)$$

or

$$\{1, 2, 2, d\}, \quad \{1, B, C, D\}, \quad 3 \leq d, \; 3 \leq B \leq C < D. \quad (2)$$

In (1), $d \geq 4$, $D \geq 4$, and $d + D = 8$; thus $d = D = 4$ and, in this case, we have the standard tetrahedral dice.

In (2), $D \geq 4$. As $8 - D = d \geq 3$ we also see that $D \leq 5$; thus D is 4 or 5. If $D = 4$ then $d = 4$ and $B = C = 3$, and we can easily check that this is not a solution. Thus, in (2) we must have $d = 3$ and $D = 5$. It is now easy to see that we must also have $B = C = 3$, and hence, in case (2), the only possible solution is to label the dice $\{1, 2, 2, 3\}$ and $\{1, 3, 3, 5\}$. To complete our solution *it is necessary to check that these labels give the correct totals and frequency distribution*, as our discussion has been based on the (possibly false) assumption that a non-standard solution exists. In fact, we have found an alternative labelling of the dice for which the sum has the same frequency distribution as for the standard dice.

- *Use a computer to check that these are the only two solutions to the problem.*

- *Use a computer to find all solutions to the problem when the labels can now be positive integers or zero.*

Let us now consider the same question for a pair of cubic dice. We could try the same method but the number of possibilities would probably be too large. We need *to describe our problem in mathematical terms*, and then use mathematics to solve it. We shall illustrate the idea by giving an alternative solution to the problem of the tetrahedral dice. The standard tetrahedral die has faces labelled $1, 2, 3, 4$, but let us now change the labels to t, t^2, t^3 and t^4, and form the function $t + t^2 + t^3 + t^4$. Let X be the score on the first die, and Y be the score on the second die; then the frequency function for the sum $X + Y$ is the function

$$(t + t^2 + t^3 + t^4)^2 = t^2 + 2t^3 + 3t^4 + 4t^5 + 3t^6 + 2t^7 + t^8$$

in the following sense. The possible sums of the faces are $2, 3$, $4, 5, 6, 7, 8$, and these occur as the indices in the polynomial; the frequencies of these sums, namely $1, 2, 3, 4, 3, 2, 1$, are the respective coefficients in the polynomial. For example, there are three ways that the sum can be 4, namely $1 + 3$, $2 + 2$ and $3 + 1$, and these correspond exactly to the number of ways that t^4 occurs on the right-hand side, namely as the products t^1t^3, t^2t^2 and t^3t^1 of terms, one from each polynomial.

To find two non-standard tetrahedral dice with the same sums, and the same distribution of sums, as a pair of standard tetrahedral dice, we label the first die with the positive integers a_1, a_2, a_3 and a_4, and the second die with the positive integers b_1, b_2, b_3, and b_4. Let A denote the score on the first die, and let B denote the score on the second die. Then the sum $A + B$ has the frequency function

$$\left(t^{a_1} + t^{a_2} + t^{a_3} + t^{a_4}\right)\left(t^{b_1} + t^{b_2} + t^{b_3} + t^{b_4}\right).$$

Thus in order to solve our problem we must find positive integers a_i and b_j such that

$$\left(t^{a_1} + t^{a_2} + t^{a_3} + t^{a_4}\right)\left(t^{b_1} + t^{b_2} + t^{b_3} + t^{b_4}\right) = (t + t^2 + t^3 + t^4)^2.$$

Let

$$P(t) = t^{a_1} + t^{a_2} + t^{a_3} + t^{a_4},$$
$$Q(t) = t^{b_1} + t^{b_2} + t^{b_3} + t^{b_4};$$

then we require

$$P(t)Q(t) = (t + t^2 + t^3 + t^4)^2 = t^2(1 + t)^2(1 + t^2)^2.$$

The problem now is to allocate the six factors t, $1 + t$ and $1 + t^2$ (each of these occurs twice) to form P and Q. As $a_1 = b_1 = 1$ (as before), each of P and Q must have a factor t. As we must also have $P(1) = Q(1) = 4$, it is easy to see that there are only two possibilities for P and Q, namely

$$P(t) = Q(t) = t(1 + t)(1 + t^2),$$

(which gives the standard tetrahedral dice) and

$$P(t) = t(1 + t)^2 = t + 2t^2 + t^3, \quad Q(t) = t(1 + t^2)^2 = t + 2t^3 + t^5,$$

which means the dice are labelled $\{1, 2, 2, 3\}$ and $\{1, 3, 3, 5\}$.

- *Solve the problem, this time without using a computer, for a pair of tetrahedral dice where now the labels can be any non-negative integers.*

- *Solve the problem of relabelling a pair of cubic dice with non–negative integers.*

9

Problem E: Solution

It is usually best to try some simple cases first. The case of two lights is trivial, for we can use switch 1 to change light 2, and switch 2 to change light 1. Now try the cases of three lights, and of four lights, to gain further insight into the problem. It seems that if $n = 3$ it is impossible to change all of the lights from ON to OFF, while it is possible to do this when $n = 4$. This suggests that the solution might depend on whether n is even or odd, so you should now try other cases to provide evidence for, or against, this suggestion. Note that to prove that it is possible to switch all of the lights OFF you only have to say which switches to use, and in which order to use them. How will you be able *to prove that it is impossible to do something*?

We should also be thinking about how to express this problem in a mathematical way. If we label the lights as $L_1, L_2, \ldots, L_n$ then we can record the state of the system as a vector $(x_1, \ldots, x_n)$, where $x_j = 0$ if the light L_j is OFF, and $x_j = 1$ if L_j is ON. For example, the vector $(0, 1, 0, 1, 1, 1)$ represents six lights $L_1, \ldots, L_6$, with L_2, L_4, L_5 and L_6 ON, and L_1 and L_3 OFF. The set S of possible states of a system with n lights is the set

$$S = \{(x_1, \ldots, x_n) : x_j = 0, 1\},$$

which has exactly 2^n elements. We shall consider the cases when n is even, and when n is odd, separately.

Case 1: n is even If we use each switch exactly once then each light will change its state $n - 1$ times, regardless of the order in which we use the switches. As $n - 1$ is odd, each light will therefore change its state. In particular, if all the lights are ON then we can switch them all OFF. Note that in this solution we use each switch once, and the switches can be used *in any order*.

It is worth noting that if we can show that the result of using several switches is independent of the order in which we use the switches, then it follows that we need use each switch at most once (for there is no point in using any switch twice). This is interesting because it implies that with n lights, there are at most 2^n combinations of switches to try, and this is exactly the same as the number of states of the lights!

Case 2: n is odd Let us say that the state $(x_1, \ldots, x_n)$ is *even* if $x_1 + \cdots + x_n$ is even, and *odd* if this sum is odd. If we use one switch we change $n - 1$ lights and, as $n - 1$ is even, this preserves the parity (even or odd) of the system. Now as n is odd, the state in which all lights are ON is an odd state, and the state in which all lights are OFF is an even state. Thus *if n is odd, and all of the lights are* ON, *then we cannot switch them all* OFF.

It is now time to set up a mathematical model for the system. We have already agreed to represent a state of the system of lights by a vector $(x_1, \ldots, x_n)$, where L_j is ON or OFF according as x_j is 1 or 0, and we have agreed to represent the set of states by $\mathcal{S}$. As each switch converts one state to another, *each switch is represented by a function* $f : \mathcal{S} \to \mathcal{S}$. We will represent switch 1 by the function f_1, and so on. As each light has only two states (ON or OFF) it is useful to work in the integers modulo 2. In fact, in this way we can express the functions f_j in the following simple form:

$$f_1(x_1, \ldots, x_n) = (x_1, \ldots, x_n) + (0, 1, 1, \ldots, 1) \quad (\text{mod } 2)$$
$$f_2(x_1, \ldots, x_n) = (x_1, \ldots, x_n) + (1, 0, 1, \ldots, 1) \quad (\text{mod } 2),$$

and so on. This is because if $x_j = 0$ then $x_j + 1 = 1$ (mod 2), and if $x_j = 1$ then $x_j + 1 = 0$ (mod 2). The notation

$$e_1 = (1, 0, \ldots, 0), \ldots, e_n = (0, \ldots, 0, 1)$$

is commonly used in the theory of vector spaces, and we shall also write $x = (x_1, \ldots, x_n)$ and $E = (1, \ldots, 1)$. Then we have the simpler formula

$$f_j(x) = x + E + e_j \quad \text{(mod 2).}$$

As

$$f_i(f_j(x)) = x + e_i + e_j \quad \text{(mod 2),}$$

it is now clear that the order in which we use the switches does not matter because, for all i, j and x, $f_i(f_j(x)) = f_j(f_i(x))$ (mod 2). One important consequence of this is that if we use many switches, each many times, and in any order, then we obtain the same effect as if we use switch 1 first for an appropriate number of times, then switch 2 for an appropriate number of times, and so on. However, as using a switch twice has no effect, it follows that *the effect of any given sequence of switches can be obtained by using switch 1 at most once, then switch 2 at most once, and so on.* Thus there are only 2^n different sequences of switches, and in each of these sequences each switch is used at most once.

It will help us familiarise ourselves with the earlier notation if we rewrite the proofs given above using the functions f_j.

Suppose that n is even. If we apply every switch once to the state $x = (x_1, \ldots, x_n)$, then the system changes to $x + (n + 1)E = x + E$ (mod 2), and this represents the statement that all of the lights change their state.

Now suppose that n is odd, and consider using only the j–th switch. As $f_j(x) = x + E + e_j$, this changes the sum of the components by the rule

$$x_1 + \cdots + x_n \mapsto x_1 + \cdots + x_n + (n + 1),$$

so that (because n is odd) the parity of the sum $\sum_j x_j$ or, equivalently, $\sum_j x_j$ (mod 2), is unchanged. This means that we cannot pass from an odd state (for example, the state in which all lights are ON), to an even state (for example, when all lights are off).

Notice that the argument in the case n is odd has proved more than we asked for: *it shows that if n is odd then we cannot pass between any even state and any odd state.*

We can obtain more information when n is even by using each of the switches $2, 3, \ldots, n$ exactly once. Then light L_1 will change its state, but none of the other lights will change (they will actually change $n - 2$

times, and $n - 2$ is even). Thus the effect of using this combination of switches is the same as switching light L_1, *and only this light*, between ON and OFF (in other words, it has the same effect as a normal switch). A similar statement is true for the other lights, so we have now proved a much stronger result which is worth stating as a theorem.

Theorem *If n is even then we can pass from any given state of the lights to any other state.*

To prove this result using the notation above, we simply note that using the switches on the lights $L_2, \ldots, L_n$ has the effect:

$$x \mapsto x + (n - 1)E + (e_2 + \cdots + e_n) = x + e_1 \quad (\text{mod } 2),$$

which is the same as applying a normal switch to L_1. In fact, we can even give a rule to show how, when n is even, we can pass from a state A to a state B. Suppose that

$$A = (a_1, \ldots, a_n), \quad B = (b_1, \ldots, b_n)$$

are two states of the system, and let

$$J = \{m : a_m = b_m\}, \quad K = \{m : a_m \neq b_m\}.$$

Of course, J and K are disjoint, $J \cup K = \{1, 2, \ldots, n\}$, and (because n is even), either both J and K have an even number of elements, or they both have an odd number of elements. We use $|J|$ and $|K|$ to denote the numbers of elements in J and K, respectively.

(1) Suppose first that $|J|$ and $|K|$ are even, and that we use all of the switches f_k, where $k \in K$, and only these. This forces the change

$$x \mapsto x + |K|E + \sum_{k \in K} e_k = x + \sum_{k \in K} e_k \quad (\text{mod } 2),$$

so only the lights L_j change, where $j \in K$. This means that A changes to B.

(2) Suppose now that $|J|$ and $|K|$ are odd, and that we use all of the switches f_j, where $j \in J$, and only these. This forces the change

$$x \mapsto x + |J|E + \sum_{j \in J} e_j = x + E + \sum_{j \in J} e_j$$

$$= x + \sum_{j \in K} e_j \quad (\text{mod } 2),$$

so again, only the lights L_j change, where $j \in K$. Again, A changes to B.

In view of our results so far, an obvious question arises.

- *If n is odd, can we pass from any even state to any other even state, and from any odd state to any other odd state?*

There are many generalisations of this problem which involve other ways to wire the lights together. For example, we can consider the situation in which the lights are arranged around a circle, and *each switch changes the state of the two lights next to it but no others.* Another possibility is that *we place lights at the vertices of a cube, and the switch on the light at a vertex v changes the state of the lights at the ends of the edges that leave v but no other lights.* In fact, we can state a very general version of the problem as follows. Let G be an abstract graph with a light at each vertex. The *neighbours* of a given vertex v are those vertices in the graph that are joined to v by an edge. We suppose that the switch attached to v changes the state of the neighbouring lights, and only these lights. The original problem is the case when the graph G is the complete graph on n vertices: this is the graph in which every pair of distinct vertices is joined by an edge.

- *Consider the case where the lights are at the vertices of a cube as labelled in Figure 9 (so that switch 1 changes L_2, L_4 and L_6, and so on). If all of the lights are on, can we switch them all off?*

Finally, the following problem suggests how we might bring more mathematics into the solution.

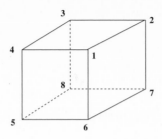

Fig. 9. Lights on a cube

- *Show that the functions $f_j : S \to S$ generate a group under composition. How can group theory now help us solve problems in this area?*

10

Problem F: Solution

We consider covering a $p \times p$ board with k triominoes and one monomino. If such a covering exists then $3k + 1 = p^2$, so that p is not a multiple of 3. Thus *if $p \equiv 0$ (mod 3) then a covering does not exist.*

Next, it is easy to see that if we can cover a $p \times p$ board with one monomino and k triominoes, then we can also cover a $(p + 3) \times (p + 3)$ board with one monomino and $k + (2p + 3)$ triominoes. Indeed, we can cover a $p \times p$ corner of the board with the single monomino and k triominoes, and then we can cover the rest of the board using $2p + 3$ more triominoes. As we can cover a 1×1 board (with one monomino and no triominoes), it follows that we can cover a $p \times p$ board whenever $p \equiv 1$ (mod 3).

The remaining case is when $p \equiv 2$ (mod 3). Clearly, we cannot cover a 2×2 board. We ask the reader to show that it is possible to cover a 5×5 board with one monomino and eight triominoes. We have now solved our problem, which we record as a theorem.

Theorem *A $p \times p$ board can be covered with one monomino and some triominoes if and only if $p \neq 2$ and p is not a multiple of 3.*

While the solution given above is elementary, it does not give us any extra information, and it does not easily generalise. In general, *the simpler the solution, the less information it gives!* We shall now give an alternative discussion which yields much more information.

First, we attach to each square of the board a label $x^a y^b$. The values of a and b are $0, 1, \ldots, p - 1$, and the labels are attached in the natural way so that the bottom row has labels (from left to right) $x^0 y^0, x^1 y^0, \ldots, x^{p-1} y^0$, and the top row has labels (from left to right) $x^0 y^{p-1}, x^1 y^{p-1}, \ldots, x^{p-1} y^{p-1}$. Note that the sum of the labels taken over all of the squares is the polynomial

$$F(x, y) = (1 + x + x^2 + \cdots + x^{p-1})(1 + y + y^2 + \cdots + y^{p-1})$$

in two variables.

Let us now assume that the board can be covered, and suppose that the single monomino is at the square with label $x^a y^b$. Each horizontal triomino occupies three adjacent squares whose labels are $x^u y^v, x^{u+1} y^v, x^{u+2} y^v$, say, and the sum of these labels is $x^u y^v (1 + x + x^2)$. Thus the sum of the labels covered by the horizontal triominoes is $g(x, y)(1 + x + x^2)$ for some polynomial g. Similarly, the sum of the labels covered by the vertical triominoes must be of the form $h(x, y)(1 + y + y^2)$ for some polynomial h. Thus, if we have covered the board as required, then we have

$$F(x, y) = x^a y^b + g(x, y)(1 + x + x^2) + h(x, y)(1 + y + y^2). \quad (1)$$

This identity (in the two variables x and y, *which may be real or complex*) is *a necessary condition for a covering to exist*. We can put (1) in a slightly more convenient form if we multiply both sides by $(1 - x)(1 - y)$; this gives the identity

$$(1 - x^p)(1 - y^p)$$
$$= (1 - x)(1 - y)x^a y^b + g(x, y)(1 - x^3)$$
$$+ h(x, y)(1 - y^3). \quad (2)$$

This algebraic approach tells us (again) that there is no solution when $p \equiv 0 \pmod 3$, for if there is we can put

$$x = y = \omega = e^{2\pi i/3}$$

in (2) and find that $(1 - \omega)^2 \omega^{a+b} = 0$ which is false.

This method enables us to answer the earlier question about the position of the monomino when $p = 5$. First, we put $x = y = \omega$ in (2) with $p = 5$ and obtain

$$\omega^{a+b} = (1 + \omega)^2 = (-\omega^2)^2 = \omega.$$

Thus $a + b \equiv 1$ (mod 3). A similar argument with $x = \omega$ and $y = \omega^2$ gives $a + 2b \equiv 0$ (mod 3). Thus

$$b = (a + 2b) - (a + b) \equiv -1 \equiv 2 \quad (\text{mod } 3).$$

We deduce that $a = b = 2$, so that *the only way to cover a 5×5 board with one monomino and eight triominoes is to put the monomino at the centre of the board.*

- *Use this method to show that the only way to cover a 4×4 board is to put the monomino in a corner. More generally, obtain further information about the possible positions of the monomino in each of the cases $p \equiv 1$ (mod 3) and $p \equiv 2$ (mod 3).*

11

Problem G: Solution

For brevity we shall often omit the word 'grams' in this discussion. Suppose that we can weigh the amounts $1, 2, \ldots, 7$ with the weights $a_1, \ldots, a_k$, where we may assume that $1 \leq a_1 \leq \cdots \leq a_k$. We must have $a_1 = 1$, else we cannot weigh one gram of the chemical. If $a_2 \geq 3$ then we cannot weigh two grams; thus a_2 is 1 or 2. Now $a_1 = 1$ and $a_2 = 2$ is probably (but not certainly) a better choice than $a_1 = a_2 = 1$, for the first choice allows us to weigh 1, 2 and 3 grams, whereas the second choice only gives us 1 and 2 grams. Thus we shall try for a solution with $a_2 = 2$. We can now weigh 1, 2 and 3 grams using only a_1 and a_2. If we take $a_3 = 4$, we can then weigh $1, 2, 3, 4, 4 + 1, 4 + 2, 4 + 3$ grams and we are done. Thus if $N = 7$ then one possible set of weights is $1, 2, 4$. *Is this the smallest number of weights and, if it is, is this the only possible set of three weights?*

Suppose that the weights $a_1, \ldots, a_k$ allow us to weigh each of the amounts $1, 2, \ldots, 7$. Then the amounts that we can weigh are the numbers $\varepsilon_1 a_1 + \cdots + \varepsilon_k a_k$, where each ε_j is 0 or 1, and not all are 0. This allows us to weigh at most $2^k - 1$ different positive amounts, so we need $2^k - 1 \geq 7$. Thus $k \geq 3$, so that *we shall always need at least three weights*. Now suppose that a_1, a_2, a_3 is *any* set of weights which allows us to weigh each amount $1, 2, \ldots, 7$. We may assume that $a_1 \leq a_2 \leq a_3$, and $a_1 = 1$. The possible amounts we can weigh are $\varepsilon_1 a_1 + \varepsilon_2 a_2 + \varepsilon_3 a_3$, where each ε_j is 0 or 1, and with these we can weigh at most seven different positive weights. If $a_i = a_j$, with $i \neq j$, then we will not be able to weigh seven *different* amounts, so it follows that we must have $1 = a_1 < a_2 < a_3$. It is now easy to see that we must have $a_2 = 2$ and $a_3 = 4$. Thus the set $\{1, 2, 4\}$ is a minimal set of weights, and it is the only minimal set. These ideas suggest the following result.

Theorem *Weights of* $1, 2, 2^2, 2^3, \ldots, 2^{k-1}$ *grams enable us to weigh any of the amounts* $1, 2, 3, \ldots, 2^k - 1$ *grams. Moreover, no other set of k weights will enable us to do this.*

Try to prove this before you read the proof below.

Proof Let $N = 2^k - 1$. Then each integer X with $0 \leq X \leq N$ can be written as a binary number, say

$$X = b_0 2^0 + b_1 2^1 + b_2 2^2 + b_3 2^3 + \cdots + b_{k-1} 2^{k-1},$$

where each b_j is 0 or 1. This tells me that the given set of weights is sufficient, for if I want to weigh X grams of the chemical, then I can do this by using precisely those weights 2^p for which $b_p = 1$. For example, if I want to weigh 109 grams, I write 109 in binary form as

$$109 = 1.2^0 + 0.2^1 + 1.2^2 + 1.2^3 + 0.2^4 + 1.2^5 + 1.2^6,$$

so that I should use the weights 1, 4, 8, 32 and 64.

The question of uniqueness is often treated lightly (or ignored) in discussions of this topic, even though it is known that if we only want to weigh amounts of $1, 2, \ldots, N$, then the set of weights may not be unique if N is not of the form $2^k - 1$. For example, each set of weights $\{1, 2, 4, 6, 9, 16\}$ and $\{1, 2, 4, 8, 16, 32\}$ will allow us to weigh any amount up to and including 40.

Suppose now that the weights $a_1, \ldots, a_p$ allow us to weigh any of the amounts $1, 2, 3, \ldots, 2^k - 1$. These p weights enable us to weigh at most $2^p - 1$ different amounts, so that $2^p - 1 \geq 2^k - 1$, and hence $p \geq k$. Moreover, if $p = k$ then we can only achieve $2^k - 1$ *different* weights if the a_j are distinct. Thus $p \geq k$ and $1 = a_1 < \cdots < a_p$. It is clear that $a_1 = 1$. As $a_2 \geq 2$, and we must be able to weigh an amount of 2, we must have $a_2 = 2$. Similarly, we must have $a_3 = 4$, and so on.   □

We now introduce a modification of this problem. We are now allowed to place each weight on either side of the scales; for example, to weigh 4 grams of the chemical I am allowed to place a weight of

6 grams on the left side of the scales, and the chemical *and* a weight of 2 grams on the right side. In the original problem we had two choices for each weight (either we use the weight or not). Here we have three choices for each weight (to use it on the left, or on the right, or not at all).

- *What is the smallest number of weights that I need in this problem, and what are these weights?*

12

Problem H: Solution

The segment $L(k)$ meets the square $S(a, b)$ if and only if the bottom left-hand corner of $S(a, b)$ lies on or below the line $x + y = k$, and the top right-hand corner lies on or above this line. Thus $L(k)$ meets $S(a, b)$ if and only if

$$a + b \leq k \leq (a + 1) + (b + 1).$$

It follows that $N(k)$ is the number of pairs (a, b) of non-negative integers such that $k - 2 \leq a + b \leq k$, and it is apparent that $N(k)$ depends critically on whether k is an integer or not. We let $[k]$ be the integer part of k.

Lemma *Let $N(k)$ be defined as above. Then*

$$N(k) = \begin{cases} 1 & \text{if } k = 0; \\ 3k & \text{if } k \text{ is a positive integer;} \\ 2[k] + 1 & \text{if } k > 0 \text{ but is not an integer.} \end{cases}$$

Proof It is clear that $N(0) = 1$. For each integer m, let $\Phi(m)$ be the number of pairs (a, b) of non–negative integers a and b with $a + b = m$. Then a simple count shows that $\Phi(m)$ is $\max\{m + 1, 0\}$.

If k is a positive integer then $\mathbb{Z} \cap [k - 2, k] = \{k - 2, k - 1, k\}$, so that

$$N(k) = \Phi(k - 2) + \Phi(k - 1) + \Phi(k)$$
$$= 3k.$$

If $k > 0$, but is not an integer, then $\mathbb{Z} \cap [k - 2, k] = \{[k - 1], [k]\}$, so that

$$
\begin{aligned}
N(k) &= \Phi([k - 1]) + \Phi([k]) \\
&= \max\{[k], 0\} + \max\{[k] + 1, 0\} \\
&= 2[k] + 1
\end{aligned}
$$

as required. $\qquad\qquad\qquad\qquad\qquad\qquad\qquad\qquad\qquad\quad$ □

We now consider the average value of $\ell(k; a, b)$ taken over those squares $S(a, b)$ that meet the segment $L(k)$. As the sum of the lengths $\ell(k; a, b)$ of the segments is the length of $L(k)$, which is $k\sqrt{2}$, we see that

$$
A(k) = \begin{cases} \sqrt{2}/3 & \text{if } k \text{ is a positive integer;} \\ \frac{k\sqrt{2}}{2[k]+1} & \text{if } k > 0, \text{ but not an integer.} \end{cases}
$$

If $k \to \infty$ through integer values, then $A(k)$ is equal to, and so tends to, $\sqrt{2}/3$. If $k \to \infty$ through non–integer values, then $k/[k] \to 1$ so that $A(k)$ tends to $1/\sqrt{2}$.

- *Can you give a simple intuitive reason why the limit is $\sqrt{2}/3$ when $k \to \infty$ through integer values?*

- *Suppose that t is chosen at random from the interval $[0, 2]$, and let $\ell_0(t)$ be the length of the intersection of the square $S(0, 0)$ with the line $x + y = t$. What is the expected value of $\ell_0(t)$, and how does this compare with the limiting values of $A(k)$ as $k \to \infty$?*

- *Now try the same problem in three dimensions. Divide the first octant of three-dimensional space into unit cubes, and consider the triangle $T(k)$ lying in the plane $x + y + z = k$, with vertices $(k, 0, 0)$, $(0, k, 0)$ and $(0, 0, k)$. What is the limiting behaviour of the average area of the intersection of $T(k)$ with a unit cube?*

13

Problem I: Solution

We can systematically list all (unordered) Diophantine pairs by writing down all factorisations of the numbers $n^2 - 1$, where $n = 2, 3, \ldots$. This list begins as follows:

$$(1, 3), \quad (1, 8), (2, 4), \quad (1, 15), (3, 5)$$
$$(1, 24), (2, 12), (3, 8), (4, 6), \quad (1, 35), \ldots$$

Given any positive integer p, we have $p(p + 2) + 1 = (1 + p)^2$ so that $(p, p + 2)$ is a Diophantine pair. Thus every positive integer occurs in some Diophantine pair.

Now consider Diophantine triples: you should try to find some of these by using a computer. In fact, there is a systematic way to pass from a Diophantine pair (a, b) to a Diophantine triple (a, b, c), and so construct infinitely many Diophantine triples.

Theorem *If (a, b) is a Diophantine pair with $ab + 1 = q^2$ then (a, b, c) is a Diophantine triple, where $c = a + b + 2q$.*

* *Use a computer to check this for a number of cases. Then give an algebraic proof. Show, however, that not every Diophantine triple arises in this way.*

Here is another way to try to find Diophantine triples.

* *Suppose that a and b are coprime and that $ab + 1 = q^2$. Show that there is some Diophantine triple (a, b, c) if and only if there are integer solutions X and Y to the Diophantine equation $bX^2 + a = aY^2 + b$.*

You can also use this result (whether you have proved it or not) with a computer to produce many Diophantine triples. For example, try to find (many) values of p, q and r such that the triples $(1, 3, p)$, $(1, 8, q)$ and $(3, 8, r)$ are Diophantine triples.

A *Diophantine 4–tuple* is a set of four integers a_1, a_2, a_3, a_4 such that if $i \neq j$ then $a_i a_j + 1$ is a square of an integer.

• *Find some Diophantine 4–tuples.*

14

Problem J: Solution

First, we should pause and think about what we mean by a polygon as a mathematical object. A *polygon* P_n is just a sequence of points (or vectors), say $v_1, \ldots, v_n$, called the *vertices* of P_n, in the plane together with the *edges* (or line segments) joining v_1 to v_2, v_2 to $v_3, \ldots$ and, finally, v_n to v_1. We shall denote the line segment joining a to b by $[a, b]$, so that an n-gon (a polygon with n vertices) can be considered as a union of the form

$$[v_1, v_2] \cup [v_2, v_3] \cup \cdots \cup [v_n, v_1].$$

At the moment there is no need to insist that these segments do not cross each other, or that a polygon has an 'inside' and an 'outside'. *We should not make assumptions until they are forced upon us.*

There is also no need to insist that a polygon is planar (that is, it lies in a plane). Indeed, as we can define the segment $[a, b]$ that joins a to b in any vector space, we can define a polygon in any vector space. Here, we shall assume that the polygons lie in some Euclidean m-dimensional space $\mathbb{R}^m$, which is just the set of all real m-tuples $(x_1, \ldots, x_m)$.

As always, it makes sense to try special cases first. The case $n = 1$ is trivial because then the polygon is $[v_1, v_1]$, which is just the set $\{v_1\}$. Then $a_1 = (v_1 + v_1)/2 = v_1$, so *there is a unique solution when $n = 1$*, namely $v_1 = a_1$.

The case $n = 2$ is more interesting, and quite different. Here we have two vertices v_1 and v_2, two coincident edges $[v_1, v_2]$ and $[v_2, v_1]$, and two midpoints $a_1 = (v_1 + v_2)/2$ and $a_2 = (v_2 + v_1)/2$. Clearly, there is no solution unless $a_1 = a_2$. If $a_1 = a_2$ there are infinitely many solutions, for we can choose any vertex v_1 and then let $v_2 = 2a_1 - v_1$.

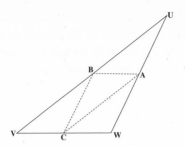

Fig. 10. Midpoints and the parallelogram

Thus *for n = 2, there may be no solutions, or there may be infinitely many solutions. In the latter case, for any point b, there is a solution with a vertex at b.*

Next, we consider $n = 3$, where the problem is as follows. *Let a_1, a_2 and a_3 be three points in the plane. Is there a triangle Δ in the plane with a_1, a_2 and a_3 the midpoints of the sides of Δ?* We should try for a solution (i) by algebra, and (ii) by geometry.

The geometric solution is easy. There are three ways to add a fourth point to a_1, a_2 and a_3 so that the four points are the vertices of a parallelogram, and these three new points are the vertices of the triangle that we are seeking. How did we reach this solution? *When we need to show that something exists, it is helpful to first assume that it does exist. Then, from this assumption, we derive further information which, we hope, can be used to provide a possible solution. Finally, we verify that this is indeed a solution (but now without making the assumption).* Let us see how this technique applies to this problem. For brevity, we use A, B and C instead of a_j.

We assume that Δ exists, and that U, V and W are its vertices (see Figure 10). It is clear that $ABCW$, $AUBC$ and $ABVC$ are parallelograms. Now, *starting only from A, B and C*, we construct the points U, V and W so that $ABCW$, $AUBC$ and $ABVC$ are parallelograms (here, we are not assuming that Δ exists). It is now easy to show that UBV, VCW and WAU are straight lines, so that there is a triangle Δ with vertices U, V and W. It is also easy to see that A, B and C are the midpoints of the sides WU, UV and VW, respectively. Incidentally, this solution shows that Δ is uniquely determined by the midpoints A, B and C.

Let us now consider an algebraic solution. We regard the points as vectors, and we use the vector a for the point A, and so on. Again we assume that Δ exists, and we let its vertices be at u, v and w. Then (up to a relabelling of the vertices) we must have

$$\tfrac{1}{2}(u + v) = b,$$
$$\tfrac{1}{2}(v + w) = c,$$
$$\tfrac{1}{2}(w + u) = a.$$

The converse is clearly also true, so these simultaneous equations are necessary and sufficient for the existence of Δ. The solutions u, v and w are obvious, *and unique*. Moreover, the vector interpretation of, for example, $u = a + (b - c)$ is evident from Figure 10. We have now solved the problem when $n = 1, 2, 3$. What would you expect when $n = 4$? Can you now make a conjecture?

Let us now consider the general problem of n points $a_1, \ldots, a_n$ which are not necessarily coplanar. Then an n-gon Δ_n exists with a_j as midpoints of its sides, and v_j as its vertices, if and only if

$$v_j + v_{j+1} = 2a_j \ (j = 1, \ldots, n), \quad v_{n+1} = v_1.$$

We ask the reader to show that *this system of equations has a unique solution if and only if n is odd*. Thus, for example, five general points in $\mathbb{R}^m$ are the midpoints of the sides of some (unique) pentagon, but six general points may or may not be the midpoints of the sides of a hexagon. Even if they are, the hexagon need not be unique.

- *Find a necessary and sufficient condition for four given points in the plane to be the midpoints of the sides of some quadrilateral. Show that if this condition holds then infinitely many such quadrilaterals exist. Is it then true that any point in the plane can be a vertex of some such quadrilateral?*

15

Problem K: Solution

Obviously, we should begin by obtaining the values of X_n from a computer. The answers will depend on the computer used but, in any case, *they are not exact*. One computer gave the following values for $n = 1, \ldots, 7$ (and the value 0 thereafter):

$$0.0501256289338006$$
$$0.00500012500624791$$
$$0.000500000125043698$$
$$0.0000500000005558832$$
$$0.000004999994416721 17$$
$$0.000000500003807246685$$
$$0.0000000502914190292358$$

This suggests that $X_n \to 0$ as $n \to \infty$, so we should prove this first. The Taylor series for $\sqrt{1-x}$ gives, for small x, $\sqrt{1-x} \sim 1 - x/2$, so we see that, with $x = 1/10^{2n}$,

$$X_n = 10^n \left(1 - \sqrt{1-x}\right) \sim 0.0 \cdots 05,$$

where there are n zeros between the decimal point and the digit 5. However, this is only an approximation. The formula for the difference of two squares shows that

$$X_n = \frac{1}{10^n + \sqrt{10^{2n} - 1}}, \tag{1}$$

and this *proves* that $X_n \to 0$ as $n \to \infty$. Notice that this also shows that $X_n > 1/(2 \times 10^n)$, so that the entry corresponding to $n = 5$ in the list is definitely incorrect. The values of X_n given by (1) were computed, and we obtained the following values:

$$0.0501256289338004$$
$$0.00500012500625039$$
$$0.000500000125000063$$
$$0.0000500000001250000$$
$$0.00000500000000012500$$
$$0.000000500000000000125$$
$$0.0000000050000000000000001$$

This list exhibits a greater degree of regularity than the first list and, from this information, we conjecture that the decimal expansion of X_n begins with n zeros, followed by 5, and then $2n - 1$ more zeros.

- *Is this conjecture correct?*

Part IV
Discussion and Generalisations

16

Problem A: Discussion and Generalisations

From now on, we shall only consider cases in which each circle lies outside the other circles. We begin by examining the problem when the triangle has been replaced by a quadrilateral. First, *we must define the problem*: we take any quadrilateral and ask whether we can find four circles, with their centres located at the vertices, such that each circle is tangent to each of the two adjacent circles. Let the lengths of the sides of the quadrilateral be L_1, L_2, L_3, L_4. Then the circles exist if and only if there is a *positive* solution to the equations

$$\begin{aligned} r_1 + r_2 &= L_3, \\ r_2 + r_3 &= L_4, \\ r_3 + r_4 &= L_1, \\ r_4 + r_1 &= L_2. \end{aligned} \tag{1}$$

If we write these equations in matrix form with M the matrix of coefficients, then we find that M has determinant zero (check this). Then linear algebra tells us that either (i) there is no solution, or (ii) there are infinitely many (but not necessarily positive) solutions. If we consider a square, and also a long thin rectangle, we can easily see that each of (i) and (ii) can arise. It is clear from (1) that $L_1 + L_3 = L_2 + L_4$ is a necessary condition for a solution to exist. Could this also be a sufficient condition for a solution to exist? We prove that it is.

Theorem *A solution exists for a quadrilateral with sides of lengths L_1, L_2, L_3 and L_4 if and only if $L_1 + L_3 = L_2 + L_4$. Moreover, if this is so, then there are infinitely many solutions to the problem.*

Proof We suppose that $L_1 + L_3 = L_2 + L_4$, and we may assume (by relabelling) that L_4 is the longest side. Now choose any r_1 and put $r_2 = L_3 - r_1$, $r_3 = L_4 - L_3 + r_1$ and $r_4 = L_2 - r_1$. It is easy to check that these r_j satisfy the equations (1), and that, providing that r_1 is sufficiently small and positive, each r_j is positive.   □

If we use the same method as above in the case of an n–gon, and label the sides $L_1, L_2, \ldots$ appropriately, then we obtain the system of equations

$$\begin{pmatrix} 1 & 1 & 0 & 0 & \cdots & 0 \\ 0 & 1 & 1 & 0 & \cdots & 0 \\ \vdots & \vdots & \vdots & \vdots & & \vdots \\ 1 & 0 & 0 & 0 & \cdots & 1 \end{pmatrix} \begin{pmatrix} r_1 \\ r_2 \\ \vdots \\ r_n \end{pmatrix} = \begin{pmatrix} L_1 \\ L_2 \\ \vdots \\ L_n \end{pmatrix},$$

where we seek positive solutions for the r_j. Let A_n be the $n \times n$ matrix in this equation. If we expand the determinant by the first column we see that

$$\det(A_n) = 1 - (-1)^n = \begin{cases} 2 & \text{if } n \text{ is odd;} \\ 0 & \text{if } n \text{ is even,} \end{cases}$$

so (from linear algebra alone)

  (i) when n is odd there is a unique solution for the r_j, and
  (ii) when n is even there is either no solution, or infinitely many
      solutions.

Suppose that n is odd. By linear algebra there is a unique solution $(r_1, \ldots, r_n)$ to the simultaneous equations. Can you find this solution and show that each r_j is positive? In the case when $n = 4$ we found a quadrilateral for which there was no solution, and another quadrilateral for which there were infinitely many positive solutions. Is the same true for any even n?

The argument given above for a quadrilateral generalises to show that if n is even, say $n = 2m$, then

$$L_1 + L_3 + \cdots + L_{2m-1} = L_2 + L_4 + \cdots + L_{2m}$$

is a *necessary condition* for a solution to exist in the case of an n–gon. Could this also be a sufficient condition?

Now consider the tetrahedron illustrated in Figure 8, and suppose that its vertices are v_1, v_2, v_3, v_4. Let the length of the side joining the vertices v_i and v_j be L_{ij}. We want to construct spheres with centres at v_j and having radius r_j so that any two of these spheres are tangent to each other. This means that we need positive solutions r_j to six equations (which you should find) in four unknowns. Thus we expect that, in general, there will be no solution (whether positive or not) of this system of equations. However, there is another, more geometric, way to analyse this case. Take a horizontal triangle in $\mathbb{R}^3$ with vertices v_1, v_2 and v_3 (the base of the tetrahedron), and construct the three spheres with centres at v_1, v_2 and v_3 such that each touches the other two (these exist, and are unique, for we have solved the problem for a triangle). For each positive r construct a sphere of radius r and place it (in $\mathbb{R}^3$, and above the plane of the triangle) so that it touches the three spheres just constructed. From physical considerations it is clear that there is exactly one position for the centre $c(r)$ of this fourth sphere. It is clear that as r varies, the point $c(r)$ moves along a curve. Thus, given v_1, v_2 and v_3, if v_4 is on this curve then our problem has a solution; if v_4 is not on this curve then our problem does not have a solution. This certainly shows that in the generic case there is no solution to the problem but in special cases there is a unique solution.

Finally, let us formulate the analogous problem for the analogue of a tetrahedron in n dimensions (for any n). The analogue of a tetrahedron in $\mathbb{R}^n$ is a set of $n+1$ points such that each pair of points is joined by an edge (think about the cases $n = 2$ and $n = 3$). Thus the general problem in $\mathbb{R}^n$ gives rise to $\frac{1}{2}n(n+1)$ equations in $n+1$ unknowns (three unknowns in two dimensions, four unknowns in three dimensions, and so on). Note that if a solution exists in n dimensions then, by restricting our attention to a 'face' of the solid, there must also be a solution for the problem with this face in dimension $n-1$. Thus, for example, there will be many solids of this type in four dimensions with no solutions.

17

Problem B: Discussion and Generalisations

First, we consider how many digits $a^{(b^c)}$ has. An integer N has d digits if and only if $10^{d-1} \le N < 10^d$ or, equivalently, $\log_{10} N < d \le 1 + \log_{10} N$. Thus N has d digits, where d is the integer part of $1 + \log_{10} N$. In particular, $a^{(b^c)}$ has d digits, where d is the integer part of $1 + (b^c) \log_{10} a$. For example, $6^{(6^6)}$ has 36306 digits.

In Part III we found the sequence of last digits of the towers of x when $x = 0, 1, \ldots, 9$, and we did this without using any significant results from number theory. However, *the more mathematics we know, the more likely we are to make progress*. Since the basic problem is to find the numbers x_n (mod 10), the following theorem (with $m_1 = 2$ and $m_2 = 5$) is bound to be useful.

Theorem (The Chinese Remainder Theorem) *Suppose that m_1 and m_2 are coprime. Then, for any u and v, the simultaneous congruences $x \equiv u \pmod{m_1}$ and $x \equiv v \pmod{m_2}$ have a unique solution $x(u, v)$ modulo $m_1 m_2$.*

We shall take $m_1 = 2$ and $m_2 = 5$, and require that $u \in \{0, 1\}$, $v \in \{0, 1, 2, 3, 4\}$ and $x(u, v) \in \{0, 1, \ldots, 9\}$. Then $x(u, v)$ is given in the following tabular form:

u	0	0	0	0	0	1	1	1	1	1
v	0	1	2	3	4	0	1	2	3	4
$x(u, v)$	0	6	2	8	4	5	1	7	3	9

For example, the column with entries 1, 2 and 7 shows that if $u = 1$ and $v = 2$ then $x(u, v) = 7$. Equivalently, this shows that, for any integer x, if $x \equiv 1$ (mod 2) and $x \equiv 2$ (mod 5), then $x \equiv 7$ (mod 10). In effect, this result reduces the problem to finding u and v, where

$$x_n \equiv u \quad (\text{mod } 2),$$
$$x_n \equiv v \quad (\text{mod } 5),$$

for then $x(u, v)$ is the last digit of x_n.

In fact, we have already solved the first of these congruences. We know that for every n, x_n is odd or even according as x is odd or even so that, for all x and all n,

$$x_n \equiv x \quad (\text{mod } 2).$$

Thus $u = 0$ if x is even, and $u = 1$ if x is odd.

If x is a multiple of 10 (that is, an even multiple of 5) then obviously x_n has last digit 0; this corresponds to the first column ($u = v = 0$) in the table. If x is an odd multiple of 5 then $u = 1$ and $v = 0$ so, as expected, $x(u, v) = 5$ and x_n has last digit 5. From now on we may assume that x is not a multiple of 5, so that x *and* 5 *are coprime*. We can now appeal to the next result (with $p = 5$) to obtain information about congruences modulo 5.

Theorem (Fermat's Theorem) *Suppose that p is prime, and x is not a multiple of p. Then $x^{p-1} \equiv 1$ (mod p).*

We conclude that if x is a positive integer that is not a multiple of 5, then $x^4 \equiv 1$ (mod 5). *At all stages, it is useful to test what we know against a specific example*, so let us do this now.

Example Let $x = 147$. Each x_n is odd so that $x_n \equiv 1$ (mod 2). As x is not a multiple of 5, $x^4 \equiv 1$ (mod 5). Also, $x \equiv -1$ (mod 4), so that

$$x_{n+1} = x^{x_n} \equiv (-1)^{x_n} = -1 \equiv 3 \quad (\text{mod } 4).$$

We conclude that $x_{n+1} = 4m + 3$, say, so that

$$x_{n+2} = 147^{4m+3} \equiv 147^3 \equiv 2^3 \equiv 3 \quad (\text{mod } 5).$$

This shows that $x_3, x_4, \ldots$ all have last digit 3. The last digit of x_1, which is 147, is 7. The last digit of x_2, which is 147^{147}, is 3.

Let us now return to the general problem. We are assuming that x is not a multiple of 5, so that $x^4 \equiv 1 \pmod 5$. Let us now write $x_n = 4a_n + b_n$, where $b_n \in \{0, 1, 2, 3\}$. Then

$$x_{n+1} = x^{x_n} = x^{4a_n + b_n} \equiv x^{b_n} \pmod 5.$$

It remains to find b_{n+1}.

First, exactly as in the example, we see that if $x \equiv 3 \pmod 4$, then $x_{n+1} \equiv 3 \pmod 4$, so that $b_{n+1} = 3$. Trivially, if $x \equiv 1 \pmod 4$, then $x_{n+1} \equiv 1 \pmod 4$, so that $b_{n+1} = 1$. Thus if x is odd, then $b_{n+1} \equiv x \pmod 4$. We conclude that if x is odd, but not a multiple of 5, then $x_{n+1} \equiv x \pmod 4$. In the remaining cases, x is even. Then x_{n+1} is an even integer to a positive even power, so that $x_{n+1} \equiv 0 \pmod 4$. These facts are worthy of a lemma.

Lemma *If x and 5 are coprime then, for all n,*

$$x_{n+1} \equiv \begin{cases} 0 & \text{if } x \text{ is even;} \\ x & \text{if } x \text{ is odd;} \end{cases} \pmod 4.$$

We can now complete our discussion for the cases in which x is not a multiple of 5.

(i) If x is even (but not a multiple of 10) then $x_{n+1} \equiv 0 \pmod 4$ so that

$$x_{n+2} = x^{x_{n+1}} = x^{4m} \equiv 1 \pmod 5.$$

As we also have $x_{n+2} \equiv 0 \pmod 2$, we see from the table on page 75 that

$$x_{n+2} \equiv 6 \pmod{10}.$$

(ii) If x is odd (and coprime with 5) then $x_{n+1} = 4m + x$, say, so that

$$x_{n+2} = x^{x_{n+1}} \equiv x^x \pmod 5.$$

We also have $x_{n+2} \equiv 1 \pmod 2$, so the last digit of x_{n+2} can now be determined directly from x^x and the table.

Note that in all cases *the towers $x_2, x_3, \ldots$ have the same last digit.*

We end with an example. Let $x = 13657853$. Then $x^4 \equiv 1 \pmod{5}$ and $x_{n+1} \equiv x \equiv 1 \pmod{4}$, so that $x_{n+1} = 4m + 1$ and

$$x_{n+2} \equiv x^{4m+1} \equiv x \equiv 3 \pmod{5}.$$

Also, $x_{n+2} \equiv 1 \pmod{2}$ so, from the table, each x_{n+2} has last digit 3.

18

Problem C: Discussion and Generalisations

It is clear that given n discs, and without identifying any patterns, there are 2^n possible starting patterns. Thus, regardless of the starting pattern, one of the patterns will eventually be repeated, and after that the patterns will occur periodically.

We need a mathematical formulation of the problem, and this is often the hardest part of problem solving. *In general, given a system in which there are exactly two choices, we should always consider using arithmetic modulo 2.* Let us now represent the red discs by 0, and the blue discs by 1. Then any state of the system is a vector of length n with each component 0 or 1, and where two vectors are considered to be the same if one can be cyclically permuted into the other. With this notation, the rules for replacement become: between 0 and 1, and between 1 and 0, we put 1; between two 0s, and between two 1s, we put 0. We can denote these rules symbolically by

$$(0, 0) \to 0, \quad (0, 1) \to 1, \quad (1, 0) \to 1, \quad (1, 1) \to 0.$$

It should now be clear that the move from one pattern to the next is given by the map

$$(x_1, \ldots, x_n) \mapsto (x_1 + x_2, x_2 + x_3, \ldots, x_n + x_1) \quad (\text{mod } 2),$$

and *this is our mathematical formulation of the problem.* As this map is a linear map it can be written in matrix form, namely

$$\begin{pmatrix} x_1 \\ x_2 \\ \vdots \\ x_n \end{pmatrix} \mapsto \begin{pmatrix} x_1 + x_2 \\ x_2 + x_3 \\ \vdots \\ x_n + x_1 \end{pmatrix} = \begin{pmatrix} 1 & 1 & 0 & \cdots & 0 \\ 0 & 1 & 1 & \cdots & 0 \\ \vdots & \vdots & \vdots & \vdots & \vdots \\ 1 & 0 & 0 & \cdots & 1 \end{pmatrix} \begin{pmatrix} x_1 \\ x_2 \\ \vdots \\ x_n \end{pmatrix},$$

where again the arithmetic is modulo 2. Let us denote the matrix of coefficients by A. Then, if we start with the pattern $(x_1, \ldots, x_n)$ and apply the process m times, the resulting pattern will be

$$A^m \begin{pmatrix} x_1 \\ x_2 \\ \vdots \\ x_n \end{pmatrix} \quad \text{(mod 2)}.$$

This observation easily enables us to check many cases on a computer. For example, if we start with the pattern (R, R, B, R, B) and apply the process 13 times, the final pattern will be (B, B, B, B, R) because (using a computer)

$$\begin{pmatrix} 1 & 1 & 0 & 0 & 0 \\ 0 & 1 & 1 & 0 & 0 \\ 0 & 0 & 1 & 1 & 0 \\ 0 & 0 & 0 & 1 & 1 \\ 1 & 0 & 0 & 0 & 1 \end{pmatrix}^{13} \begin{pmatrix} 0 \\ 0 \\ 1 \\ 0 \\ 1 \end{pmatrix} = \begin{pmatrix} 1 \\ 1 \\ 1 \\ 1 \\ 0 \end{pmatrix} \quad \text{(mod 2)}.$$

A **warning** is appropriate here. If we compute A^n for a large n, and then reduce the resulting matrix modulo 2, the answer may not be correct because there may be errors in computing A^n. The correct procedure is to reduce the answer modulo 2 *after each product*.

We now prove some general results about these patterns and, for brevity, we shall use the same symbol, say x, for both a (row) vector and its transpose (a column vector). We fix n and let $\mathbf{0} = (0, \ldots, 0)$ and $\mathbf{1} = (1, \ldots, 1)$. Obviously, if we start with one of the patterns $\mathbf{0}$ or $\mathbf{1}$ then we move immediately to $\mathbf{0}$ and stay there. It is not so obvious, but true, that if we reach the pattern $\mathbf{0}$ then we must have started at $\mathbf{0}$ or $\mathbf{1}$. To show this, suppose that we have $Ax = \mathbf{0}$; then

$$x_1 + x_2 = x_2 + x_3 = \cdots = x_n + x_1 = 0 \quad \text{(mod 2)},$$

and (as each x_j is 0 or 1) this means that $x_1 = \cdots = x_n$, so that x is $\mathbf{0}$ or $\mathbf{1}$. More generally, if $A^m x = \mathbf{0}$ for some x and some m, then x is $\mathbf{0}$ or $\mathbf{1}$, and $Ax = \mathbf{0}$.

Next, we observe that for any pattern x, the next pattern Ax has an *even number* of blue discs. This is because if $y = Ax$ then

$$y_1 + \cdots + y_n = (x_1 + x_2) + \cdots + (x_n + x_1),$$

and this sum is even. We shall say that a pattern x is *odd* or *even* according as $\sum_j x_j$ is odd or even, respectively. Thus *only the first pattern can be odd*.

If the first pattern is odd, the second pattern is even, so either the first or the second pattern is even. We shall now show that *if n is odd then, starting from a first even pattern, the sequence of patterns is cyclic* (see Part III and the case $n = 5$). Let $\mathcal{E}$ be the set of even patterns. Then A maps $\mathcal{E}$ into itself, and we shall now show that $A : \mathcal{E} \to \mathcal{E}$ is injective. As $\mathcal{E}$ is a finite set this will show that A is a bijection of $\mathcal{E}$ onto itself, and so has an inverse, say A^{-1} (note that we are not claiming that A has an inverse on the set of all patterns, but only on $\mathcal{E}$). Since A is a permutation of $\mathcal{E}$, it must be of finite order or, equivalently, $A^p = I$ for some positive integer p, and this is the result we want. It remains to show that the map A of $\mathcal{E}$ into itself is injective. Suppose that $x, y \in \mathcal{E}$ and that $Ax = Ay$. Then, working modulo 2, we have $A(x + y) = Ax + Ay = 2Ax = \mathbf{0}$. This means that $x + y$ is $\mathbf{0}$ or $\mathbf{1}$. However, $x + y$ is in $\mathcal{E}$ and $\mathbf{1}$ is not (because n is odd). Thus $x + y = \mathbf{0}$, and then $x = x + (x + y) = 2x + y = y$ as required.

We have just proved a result in the case when n is odd. The cases where n is even seem more difficult to analyse, but we do have the following result which the reader can check experimentally on a computer.

Theorem *If $n = 2^m$ for some m, then eventually all discs will be red (and will stay red).*

Proof Let A be the $n \times n$ matrix that represents the change of patterns; then we have to show that if $n = 2^m$ there is some positive integer k such that $A^k = 0$ (the zero matrix). In fact, we shall show that $A^n = 0$ (mod 2). We write $A = I + C$, where, for example, if $n = 4$ we have

$$I = \begin{pmatrix} 1 & 0 & 0 & 0 \\ 0 & 1 & 0 & 0 \\ 0 & 0 & 1 & 0 \\ 0 & 0 & 0 & 1 \end{pmatrix}, \quad C = \begin{pmatrix} 0 & 1 & 0 & 0 \\ 0 & 0 & 1 & 0 \\ 0 & 0 & 0 & 1 \\ 1 & 0 & 0 & 0 \end{pmatrix}.$$

As the effect of applying the matrix C is the cyclic permutation

$$(x_1, \ldots, x_n) \mapsto (x_2, x_3, \ldots, x_n, x_1),$$

we see that $C^n = I$, where I is the identity matrix.

Next observe that, as we are working modulo 2, for any matrix X we have

$$(I + X)^2 = I + 2X + X^2 = I + X^2.$$

More generally, if $(I + X)^{2^m} = I + X^{2^m}$, then

$$(I + X)^{2^{m+1}} = (I + X^{2^m})^2 = I + X^{2^{m+1}},$$

so that (by induction), if $n = 2^m$ then $(I + X)^n = I + X^n$. Thus, working modulo 2,

$$A^n = (I + C)^n = I + C^n = I + I = 2I = 0,$$

as required. For example, if $n = 16 = 2^4$ then, after 16 changes of pattern, all of the discs will be red.   □

The fixed states (or fixed patterns) are of interest. It is clear that $(R, R, \ldots, R, R)$ is a fixed state, but there are other fixed states; for example, $(R, B, B, \ldots, R, B, B)$, where n is a multiple of 3, is a fixed state. One fact about the fixed states is immediately clear: any pattern that is fixed is the same as its next pattern, and this next pattern is even; thus *in any fixed pattern there must be an even number of blue discs.*

Finally, how many different starting patterns are there? The interested reader should consult almost any book on group theory, and find and study *Burnside's Theorem*.

19

Problem D: Discussion and Generalisations

First, we solve the problem for the tetrahedral dice, where the labels can now be *non-negative* integers. We argue exactly as before: we let

$$P(t) = t^{a_1} + t^{a_2} + t^{a_3} + t^{a_4},$$
$$Q(t) = t^{b_1} + t^{b_2} + t^{b_3} + t^{b_4},$$

where now the a_i and b_j may be zero, and we obtain the same identity, namely

$$P(x)Q(x) = x^2(1+x)^2(1+x^2)^2.$$

This gives a finite number of possibilities for P and Q, but our choices of P and Q must satisfy the constraint $P(1) = 4 = Q(1)$. With this constraint there are nine possibilities, but if we reject the cases that are obtained by interchanging P and Q from some other case, the list reduces to the following five possibilities for the pair $P(x), Q(x)$:

$$
\begin{array}{ll}
x^2 + 2x^3 + x^4, & 1 + 2x^2 + x^4, \\
x + 2x^2 + x^3, & x + 2x^3 + x^5, \\
1 + 2x + x^2, & x^2 + 2x^4 + x^6, \\
x^2 + x^3 + x^4 + x^5, & 1 + x + x^2 + x^3, \\
x + x^2 + x^3 + x^4, & x + x^2 + x^3 + x^4.
\end{array}
$$

These give the solutions as

$$\{2, 3, 3, 4\}, \qquad \{0, 2, 2, 4\},$$
$$\{1, 2, 2, 3\}, \qquad \{1, 3, 3, 5\},$$
$$\{0, 1, 1, 2\}, \qquad \{2, 4, 4, 6\},$$
$$\{2, 3, 4, 5\}, \qquad \{0, 1, 2, 3\},$$
$$\{1, 2, 3, 4\}, \qquad \{1, 2, 3, 4\}.$$

These are the only solutions to the problem. It is interesting to note that each of these solutions is obtained from one of the positive solutions $\{1, 2, 2, 3\}$ and $\{1, 3, 3, 5\}$ by adding 1 to each label on one die, and subtracting 1 from each label on the other die. In retrospect, we should have seen, and remarked, at the start that this process will give us additional solutions.

The same technique can be used to solve the problem with a pair of standard (cubic) dice with faces labelled $1, \ldots, 6$. The possible totals are $2, 3, \ldots, 12$, and the frequency distribution of these totals can be found directly from the identity

$$(x^1 + x^2 + \cdots + x^6)^2 = x^2 + 2x^3 + \cdots + 6x^7 + \cdots + 2x^{11} + x^{12}.$$

Thus, to solve the problem of relabelling a pair of cubic dice, we must find non-negative integers $a_1, \ldots, a_6$ (on the first die), and $b_1, \ldots, b_6$ (on the second die), such that

$$(x^{a_1} + \cdots + x^{a_6})(x^{b_1} + \cdots + x^{b_6}) = (x^1 + \cdots + x^6)^2$$
$$= x^2(1 + x)^2(1 + x + x^2)^2(1 - x + x^2)^2.$$

As before, we write

$$P(x) = x^{a_1} + \cdots + x^{a_6}, \qquad Q(x) = x^{b_1} + \cdots + x^{b_6},$$

and we have to find positive integers a_i and b_j such that

$$P(x)Q(x) = x^2(1 + x)^2(1 + x + x^2)^2(1 - x + x^2)^2$$

subject to the constraints $P(1) = Q(1) = 6$.

We may assume that $a_1 \leq a_2 \leq \cdots \leq a_6$, and similarly for the b_j, and we must have $a_1 = b_1 = 1$ (so that both P and Q have a factor x)

and $a_6 + b_6 = 12$. It follows that we must have

$$P(x) = x(1 + x)^r(1 + x + x^2)^s(1 - x + x^2)^t,$$
$$Q(x) = x(1 + x)^{2-r}(1 + x + x^2)^{2-s}(1 - x + x^2)^{2-t},$$

where each of r, s and t are 0, 1 or 2. Now we must also have $6 = P(1) = 2^r 3^s 1^t$, and this means that $r = s = 1$. Thus there are only two possible solutions, namely

$$P(x) = x(1 + x)(1 + x + x^2)(1 - x + x^2)^t,$$
$$Q(x) = x(1 + x)(1 + x + x^2)(1 - x + x^2)^{2-t},$$

where $t \in \{0, 1, 2\}$. The case $t = 1$ gives the standard dice. The only alternative (up to interchanging P and Q) is

$$P(x) = x(1 + x)(1 + x + x^2),$$
$$Q(x) = x(1 + x)(1 + x + x^2)(1 - x + x^2)^2.$$

We now expand these polynomials and find that

$$P(x) = x + 2x^2 + 2x^3 + x^4, \quad Q(x) = x + x^3 + x^4 + x^5 + x^6 + x^8;$$

thus *the only possible ways to label the pair of cubic dice are the standard labelling, and the labelling* 1, 2, 2, 3, 3, 4 *and* 1, 3, 4, 5, 6, 8. Again, we should check that the second labelling does give the correct distribution, which it does.

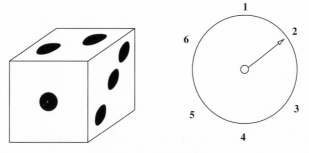

Fig. 11. A die and a pointer

This problem generalises in the following way. Throwing a die has the same effect as spinning a pointer on a disc (see Figure 11), so we can ask the same question about two (or more) pointers each spinning on a disc. In this formulation we can consider the first disc to be labelled $1, 2, \ldots, p$ and the second disc $1, 2, \ldots, q$, for any positive integral values of p and q.

This problem can be analysed using cyclotomic polynomials. Without going into details (the interested reader can look elsewhere) the cyclotomic polynomials are monic polynomials $\varphi_k(x)$, $k = 2, 3, \ldots$, with integer coefficients, which are irreducible over the integers, and which are such that $1 + x + x^2 + \cdots + x^{n-1}$ is the product of the $\varphi_d(x)$ taken over those d such that $d > 1$ and d divides n. Clearly, then, the polynomial attached to the labelling of a die with n faces must be a product of these cyclotomic polynomials.

Finally, we can ask the following question. Let X and Y be finite sets of integers with given, but possibly different, probability distributions. Choose a number x from X, and a number y from Y, each according to the given probabilities. Now find the set T of possible totals $x + y$ and the associated probability distribution. The problem now is to find other probability distributions on X and Y such that the sum $x + y$ has the same probability distribution as before. For example, we could consider tossing a fair coin and a biased coin. As an example of this situation, let $X = Y = \{1, 2\}$, with probabilities

$$P_X(1) = \tfrac{1}{3}, \quad P_X(2) = \tfrac{2}{3}, \quad P_Y(1) = P_Y(2) = \tfrac{1}{2}.$$

What values of a, b, c and d are possible such that if

$$P_X(1) = a, \quad P_X(2) = b, \quad P_Y(1) = c, \quad P_Y(2) = d,$$

then $x + y$ has the same distribution as before? The reader may wish to explore these ideas further.

20

Problem E: Discussion and Generalisations

First, the solution to the problem of the cube is easy. If we use every switch once, then each light will change three times, and hence each light will change from ON to OFF. In fact, if we operate the three switches that are adjacent to a given vertex this has the same effect as a normal switch at that vertex; thus we can pass from any one state of the lights to any other state.

Let us continue by addressing the question: if, in the original problem, n is odd, can we pass from any even state to any other even state, and from any odd state to any other odd state? It is always a good idea to experiment with some special cases, so here is one. Suppose that $n = 5$ and that $A = (1, 0, 1, 0, 1)$ and $B = (1, 1, 1, 0, 0)$: *can we pass from A to B?* Let us agree to use switch 1 k_1 times, switch 2 k_2 times, and so on, where each k_j is 0 or 1. We will pass from A to B with this choice of switches if and only if we can solve the congruence equation

$$A + \sum_{j=1}^{5} k_j(E + e_j) = B \pmod 2,$$

where the k_j are the 'unknowns' (and each is 0 or 1). This equation is equivalent to the equation

$$\sum_{j=1}^{5} k_j(E + e_j) = B + A = (0, 1, 0, 0, 1) \pmod 2,$$

or to the system

$$0 + k_2 + k_3 + k_4 + k_5 = 0,$$
$$k_1 + 0 + k_3 + k_4 + k_5 = 1,$$
$$k_1 + k_2 + 0 + k_4 + k_5 = 0,$$
$$k_1 + k_2 + k_3 + 0 + k_5 = 0,$$
$$k_1 + k_2 + k_3 + k_4 + 0 = 1,$$

taken modulo 2. By subtraction we see that $k_1 = k_3 = k_4$ and $k_2 = k_5$, and hence the unique solution (modulo 2) is $k_2 = k_5 = 1$ and $k_1 = k_3 = k_4 = 0$. Thus we should use only switches 2 and 5, and it is easy to check that this combination does indeed take us from state A to state B. More generally, if we consider any two states A and B and write $A + B = C = (c_1, c_2, c_3, c_4, c_5)$, then we obtain the equations

$$0 + k_2 + k_3 + k_4 + k_5 = c_1,$$
$$k_1 + 0 + k_3 + k_4 + k_5 = c_2,$$
$$k_1 + k_2 + 0 + k_4 + k_5 = c_3,$$
$$k_1 + k_2 + k_3 + 0 + k_5 = c_4,$$
$$k_1 + k_2 + k_3 + k_4 + 0 = c_5.$$

If we now add all of these equations together, we find that a necessary condition for a solution is that $c_1 + \cdots + c_5 = 0 \pmod 2$. This confirms that (with $n = 5$) we can only pass between even states, and between odd states.

In order to solve the general problem in the case when n is odd, we shall explore the group-theoretic aspects of this situation, and then use our additional information to solve the problem. First, we show that the f_j generate a group of permutations of S. As each f_j is a map of S into itself, we can certainly form the composition $f_i(f_j(x))$ of any two maps. Next, the composition of mappings is always associative (this is a general fact). Clearly we must include the identity map f_0 which represents the use of *none* of the switches. Finally, it is clear that using any switch twice has no effect: thus $f_j(f_j(x)) = x = f_0(x)$. Thus each f_j is a permutation of S, with each f_j being its own inverse. The set of f_j therefore generates a group (of permutations of S) which we shall denote by F. As the actions of the switches commute, F is an abelian group. This is a special case of the result that says *if every element of a group G (with identity e) has order two, then G is abelian.* The proof (for a general group) is easy: if x and y are in G, and every element of G has order two, then

$$xy = xey = x(xy)^2 y = xxyxyy = (x^2)yx(y)^2 = yx.$$

It is also known that *if every element of a finite abelian group has order two then the group has order* 2^m *for some m.* We already know this in our case because we have seen that F is simply the collection of maps $f_1^{a_1} \cdots f_n^{a_n}$, where each a_j is 0 or 1.

We can make much more progress by studying the powerful theory of groups acting on a set. Consider the most general situation in which G is some group of permutations of a set X. If $w \in X$ then the *stabiliser* of the point w is the set of g in G that fix w; that is, it is the subgroup $\{g \in G : g(w) = w\}$ of G. The *orbit* of w is the set of points in X that w can be mapped to by the elements of G; that is, it is the subset $\{g(w) : g \in G\}$ of X. We write

$$\text{stab}(w) = \{g \in G : g(w) = w\},$$
$$\text{orb}(w) = \{g(w) : g \in G\}.$$

These are extremely important ideas, and the reader is urged to learn more about them by consulting appropriate texts. For example, the following result (in which $|X|$ denotes the number of elements in a set X) is one of the most important theorems in this area.

Theorem (The Orbit-Stabiliser Theorem) *In the situation described above,* $|\text{orb}(w)| \times |\text{stab}(w)| = |G|$.

In the case we are considering here, $X = S$ (each w in X is a state of the system of lights), and G is the group generated by the switches $f_1, \ldots, f_n$. We have seen that $|G| = 2^n = |S|$, and this has an important corollary.

Corollary *If the stabiliser of some state in* S *consists only of the identity map, then the orbit of* w *is the entire set* S.

The proof is easy. If $|\text{stab}(w)| = 1$ then $|\text{orb}(w)| = |G|$, and as $|G| = |\mathcal{S}|$, we see that $|\text{orb}(w)| = |\mathcal{S}|$. In plain language this means that if there is a state w which is only left unchanged by using no switches at all, then we can pass from any state to any other state of the system by using some combination of the switches.

Let us see how this idea applies to the case of the lights on the cube illustrated in Figure 9. Let x be any state, and suppose that when we use switch 1 k_1 times, switch 2 k_2 times, and so on, we leave the state x unchanged. The only switches that affect the light L_1 are the switches at L_2, L_4 and L_6; thus, as L_1 is unchanged, we must have

$$k_2 + k_4 + k_6 = 0 \quad (\text{mod } 2).$$

Similarly, the only switches that affect the light L_3 are the switches at L_2, L_4 and L_8, so that we must have

$$k_2 + k_4 + k_8 = 0 \quad (\text{mod } 2).$$

We conclude that we must have $k_6 = k_8$ (mod 2). By symmetry we must have

$$k_2 = k_4 = k_6 = k_8 \quad (\text{mod } 2),$$

and as $k_2 + k_4 + k_6 = 0$ (mod 2), we conclude that $3k_2 = 0$ (mod 2). This tells us that k_2 is even, hence $k_2 = 0$. We conclude that $k_2 = k_4 = k_6 = k_8 = 0$ and, by symmetry again, that $k_1 = k_3 = k_5 = k_7 = 0$. In other words, we have shown that the identity map is the only permutation in the stabiliser of the state x. Thus we can conclude that *for the lights at the vertices of a cube, we can pass from any state to any other state by a suitable combination of the switches.*

We shall now use the Orbit-Stabiliser Theorem to answer a question raised earlier.

Theorem *In the general problem of switching n lights, where n is odd, we can pass between any two even states, and between any two odd states, but not between an odd and an even state.*

Proof Let $\mathcal{E}$ be the set of even states, and $\mathcal{O}$ the set of odd states. We already know that we cannot pass between an even state and an odd state, so the orbit of an even state lies entirely within $\mathcal{E}$, and the orbit of an odd state lies entirely within $\mathcal{O}$. It is sufficient, therefore, to show that there are exactly two orbits; then one orbit will have to be $\mathcal{E}$, and the other orbit will have to be $\mathcal{O}$.

Suppose now that we can show that the stabiliser of any state s is the group $\{f_0, f_1 f_2 \cdots f_n\}$. As this stabiliser has exactly two elements, the Orbit-Stabiliser Theorem implies that the orbit of s has exactly 2^{n-1} points in it. As s is any state, and as the total number of states is 2^n, this implies that there are exactly two orbits, and this is what we want to prove.

It remains, then, to show that the stabiliser of any state s is the group $\{f_0, f_1 f_2 \cdots f_n\}$. The general element of G is $f_1^{t_1} f_2^{t_2} \cdots f_n^{t_n}$, and let us suppose that this leaves the state $s = (a_1, \ldots, a_n)$ unchanged. As a_j is unchanged (modulo 2) we must have

$$(t_1 + \cdots + t_n) + t_j \equiv 0 \pmod 2.$$

But this means that

$$t_1 = t_2 = \cdots = t_n \pmod 2,$$

so the only possibilities are $t_1 = \cdots = t_n = 0$ (which gives f_0), and $t_1 = \cdots = t_n = 1$ (which gives $f_1 \cdots f_n$). The proof is complete. $\square$

Here is another approach to the problem which the reader may find easier to use when programming. We shall now use -1 for OFF and 1 for ON; then, by using matrices, we can use multiplication rather than addition modulo 2. Each switch is given by the action of a diagonal matrix (acting on the column vectors) each of whose components are 1 or -1. For example, if $n = 4$, then the function f_4 (that is, the action of switch 4) changes the states of the lights 1, 2 and 3 only, so f_4 has the following action:

$$f_4 : \begin{pmatrix} x_1 \\ x_2 \\ x_3 \\ x_4 \end{pmatrix} \mapsto \begin{pmatrix} -x_1 \\ -x_2 \\ -x_3 \\ x_4 \end{pmatrix} = \begin{pmatrix} -1 & 0 & 0 & 0 \\ 0 & -1 & 0 & 0 \\ 0 & 0 & -1 & 0 \\ 0 & 0 & 0 & 1 \end{pmatrix} \begin{pmatrix} x_1 \\ x_2 \\ x_3 \\ x_4 \end{pmatrix}.$$

It is useful to summarise what we have learnt from this exercise.

- Any finite graph, say with n vertices, gives us a 'switching-lights' problem.
- Each switch can be represented by a permutation of the set of 2^n states of the system, and also by an $n \times n$ matrix. The matrix is always a diagonal matrix with the diagonal elements ± 1. The actual form of the matrix (or permutation) depends on the nature of the original graph.
- The orbit of a state s is precisely the set of states that s can be transformed into by using the switches.
- The Orbit-Stabiliser Theorem enables us to count the number of elements in an orbit by computing the stabiliser of a state (that is, by counting those combinations of switches that leave the state unchanged).

- *Give a complete analysis of the situation when lights are placed at the vertices of a regular n-gon, and each switch changes only those two lights that lie either side of the switch.*

- *Generalise the problem for a cube to the corresponding problem in which the lights are placed at the vertices of each of the five regular solids (the remaining four solids are the tetrahedron, the octahedron, the dodecahedron and the icosahedron).*

21

Problem F: Discussion and Generalisations

We assume that a $p \times p$ board can be covered, and that the monomino is at the square labelled $x^a y^b$. Then there are polynomials $g(x, y)$ and $h(x, y)$ such that (2) in Part III holds. Again, we let $\omega = e^{2\pi i/3}$, and we consider the two cases $p \equiv 1$ and $p \equiv 2$ (mod 3).

Case 1: $p \equiv 1$ (mod 3). In this case $\omega^p = \omega$. Thus if we put $x = y = \omega$ in (2) we obtain $\omega^{a+b} = 1$. This implies that $a + b \equiv 0$ (mod 3). Next, we put $x = \omega$ and $y = \omega^2$ in (2) and obtain $a + 2b \equiv 0$ (mod 3). The only solution to these simultaneous congruences is $a \equiv b \equiv 0$ (mod 3), so that a covering of a $p \times p$ square with $p \equiv 1$ (mod 3) is only possible if the monomino is placed at a square with a label $x^{3s} y^{3t}$. In particular, if $p = 4$ then the monomino must be placed in a corner of the board. Although it is easy to give an ad hoc proof of this fact, *it is always preferable to give a proof which is likely to be generalised.*

Case 2: $p \equiv 2$ (mod 3). Again we take $x = y = \omega$ and we obtain $a + b \equiv 1$ (mod 3). If we take $x = \omega$ and $y = \omega^2$ we obtain $a + 2b \equiv 0$ (mod 3), so we find that $a \equiv b \equiv 2$ (mod 3). Thus in this case the monomino must be placed at a square labelled $x^{3s+2} y^{3t+2}$. In particular, if $p = 5$ then the monomino must be placed at the centre of the board.

- *Does a covering exist in Case 1 for any admissible choice of s and t? Does a covering exist in Case 2 for any admissible choice of s and t?*

- *Is it possible to cover a 3×7 board using one monomino and ten 2×1 tiles? Is it possible to cover a 5×5 board using one 4×1 tile and seven triominoes?*

It should be clear to the reader that a similar approach might help us obtain information about coverings of a rectangular board, and possibly with other shapes. In principle, similar problems could be tackled in three dimensions, where we could use a cuboid as a 'board', and where we would label the cubes that make up the cuboid with labels of the form $x^a y^b z^c$. If we wish to pursue this topic further, it might be advantageous to study the theory of factorisation of polynomials in several variables. On the other hand, it might not!

22

Problem G: Discussion and Generalisations

The discussion in Part III should lead us to suspect that in the new problem we should work in base three arithmetic, and use the weights $1, 3, 3^2, 3^3, \ldots$. It is easy to check that by taking the weights $1, 3, 9$ we can weigh all amounts up to and including 13 (which is $1 + 3 + 9$). If we then try with the weights $1, 3, 9$ and 27, and so on, we might make the following conjecture.

- *Conjecture: we can weigh any integral amount up to and including* $1 + 3 + 3^2 + \cdots + 3^{k-1}$, *which is* $\frac{1}{2}(3^k - 1)$, *by using only the k weights* $1, 3, 3^2, \ldots, 3^{k-1}$.

We leave the reader to explore this problem further, but we shall give some hints in the form of an example.

Example Suppose that I want to weigh 6402 grams of the chemical, and only the weights $1, 3, 3^2, 3^3, \ldots, 3^{k-1}$ are available. Can I do this, and which value of k should I use? First, note that the most that I can weigh with these weights is their sum, namely $\frac{1}{2}(3^k - 1)$, so we certainly need

$$3^k \geq (2 \times 6402) + 1 = 12805,$$

or $k \geq 9$. We shall now attempt to weigh 6402 grams of the chemical using only the weights $1, 3, 3^2, \ldots, 3^8$.

Suppose that I can weigh X grams of the chemical, where the chemical and some of the weights are on the left side of the scales, other weights are on the right side, and some weights are not used at all. Let L be the set of integers m such that the weight 3^m is on the left side of the scales, let R be the set of m such that the weight 3^m is on the right side, and let N be the set of m for which the weight 3^m is not used.

Clearly, the sets L, R and N are disjoint, and their union is $K = \{0, 1, 2, \ldots, k-1\}$, so that

$$\sum_{j=0}^{k-1} 3^j = \sum_{j \in L} 3^j + \sum_{j \in R} 3^j + \sum_{j \in N} 3^j.$$

As

$$X + \sum_{j \in L} 3^j = \sum_{j \in R} 3^j,$$

we see that

$$X + \tfrac{1}{2}(3^k - 1) = X + \sum_{j=0}^{k-1} 3^j = 2\sum_{j \in R} 3^j + \sum_{j \in N} 3^j.$$

This simply means that, for a given X, we find k as above and then express the sum $X + \frac{1}{2}(3^k - 1)$ in base three. Then, from this expression, we can find the sets L and R. For example, if $X = 6402$ we take $k = 9$, and find that $6402 + \frac{1}{2}(3^9 - 1) = 16243$, and

$$16243 = (2 \times 3^8) + 3^7 + 3^6 + (2 \times 3^4) + 3^3 + 3^2 + (2 \times 3) + 1.$$

Thus $6402 = 3^8 - 3^5 + 3^4 + 3$, so I put weights 3^8, 3^4 and 3 on one side, and the weight 3^5 with the chemical on the other side.

23

Problem H: Discussion and Generalisations

First, suppose that k is an integer. If the segment $L(k)$ meets the square $S(a, b)$ then $L(k) \cap S(a, b)$ is either a diagonal (of length $\sqrt{2}$) of $S(a, b)$, or a vertex (of length 0) of $S(a, b)$. It is easy to see that there are approximately twice as many squares $S(a, b)$ for which the intersection is a point as there are for which the intersection is the diagonal. Thus the limiting answer should be $(\sqrt{2} + 0 + 0)/3$ (which it is).

Now consider the second question in Part III. As t has a uniform probability distribution on $[0, 2]$, the appropriate probability density function is $\frac{1}{2}dt$. Next, a simple calculation gives

$$\ell_0(t) = \begin{cases} \sqrt{2}t & \text{if } 0 \leq t \leq 1; \\ \sqrt{2}(2 - t) & \text{if } 1 \leq t \leq 2. \end{cases}$$

Since it is obvious that $\ell_0(t) = \ell_0(2 - t)$ we need only verify the first formula here. However, we should also verify the second formula directly as this will serve as a check on our working. In any case, the expected value of $\ell_0(t)$ is

$$E(\ell_0) = \frac{1}{2} \int_0^2 \ell_0(t)\, dt = \frac{1}{\sqrt{2}},$$

which is the same as the limiting value of $A(k)$ as $k \to \infty$ avoiding integer values.

Now consider the three-dimensional problem. For non-negative integers a, b and c, let $C(a, b, c)$ be the cube in $\mathbb{R}^3$ that corresponds to the square $S(a, b)$ in $\mathbb{R}^2$. Also, for each integer m, let $\Psi(m)$ be the number of triples (a, b, c) of non-negative integers such that $a + b + c = m$.

Finally, suppose that $k \geq 0$, and let $n(k)$ be the number of cubes $C(a, b, c)$ that meet the triangle $T(k)$ defined in Part III.

If, for example, k is positive, non-integral and large, then $T(k)$ meets $C(a, b, c)$ if and only if

$$a + b + c \leq k \leq (a + 1) + (b + 1) + (c + 1),$$

or $k - 3 \leq a + b + c \leq k$, so that

$$n(k) = \Psi(m - 2) + \Psi(m - 1) + \Psi(m),$$

where $m = [k]$, the integer part of k. As the area of $T(k)$ is $(\sqrt{3}/2)k^2$, the average area of $T(k) \cap C(a, b, c)$ taken over those cubes that meet $T(k)$ is

$$\frac{k^2 \sqrt{3}}{2[\Psi([k] - 2) + \Psi([k] - 1) + \Psi([k])]}.$$

It only remains to find $\Psi(m)$. However, it is clear that

$$\Psi(m) = \sum_{c=0}^{m} \Phi(m - c) = \sum_{c=0}^{m} (m - c + 1) = \tfrac{1}{2}(m + 1)(m + 2).$$

24

Problem I: Discussion and Generalisations

The Theorem in Part II is easy to prove by showing that $bc + 1 = (b + q)^2$ and $ac + 1 = (a + q)^2$. Not every Diophantine triple arises in this way; for example, $(1, 3, 8)$ and $(1, 3, 120)$ are both Diophantine triples and at least one of these cannot be of this form.

Let us consider the problem of extending a Diophantine pair (a, b), where a and b are coprime and $ab + 1 = q^2$, to a Diophantine triple (a, b, c). We have to find positive integers c, X and Y such that $ac + 1 = X^2$ and $bc + 1 = Y^2$. If there is a c such that (a, b, c) is a Diophantine triple then X and Y exist and are solutions of the Diophantine equation $bX^2 - b = aY^2 - a$. Conversely, if X and Y are solutions of this equation, then a divides $X^2 - 1$ (because a and b are coprime), and b divides $Y^2 - 1$, so we can define c by $c = (X^2 - 1)/a = (Y^2 - 1)/b$, and then (a, b, c) is a Diophantine triple.

We can use this argument to make a reasonably efficient computer search for those values of c for which (a, b, c) is a Diophantine triple (by searching for solutions X and Y of $bX^2 = aY^2 + b - a$). This gives, for example, the following Diophantine triples:

$$(1, 3, 8), \ldots, (1, 3, 326040),$$

$$(1, 8, 15), \ldots, (1, 8, 609960),$$

$$(3, 8, 21), \ldots, (3, 8, 1154440).$$

Note that this shows that many Diophantine triples (a, b, c) can be found with the same a and b but with different c.

Fermat was the first person to find a Diophantine 4-tuple, namely $(1, 3, 8, 120)$, and (much later) Baker and Davenport (and subsequently others) showed that if $(1, 3, 8, d)$ is a Diophantine 4–tuple, then $d = 120$ (but this is not easy). No Diophantine 5–tuple (of integers) is known; however, $(777480/8288641, 1, 3, 8, 120)$ is a Diophantine 5–tuple of *rational numbers*.

It is easy to see that $(k - 1, k + 1, 4k)$ is a Diophantine triple for $k = 2, 3, \ldots$. This Diophantine triple extends to the Diophantine 4–tuple

$$(k - 1, k + 1, 4k, 16k^3 - 4k),$$

because of the identities

$$(k - 1)(16k^3 - 4k) + 1 = (4k^2 - 2k - 1)^2,$$
$$(k + 1)(16k^3 - 4k) + 1 = (4k^2 + 2k - 1)^2,$$
$$4k(16k^3 - 4k) + 1 = (8k^2 - 1)^2,$$

and this provides an infinite supply of Diophantine 4–tuples. The case $k = 2$ is Fermat's example.

We can construct families of Diophantine triples from the generalised Fibonacci sequences. The Fibonacci sequence F_n is defined by

$$F_0 = 0, \quad F_1 = 1, \quad F_{n+2} = F_{n+1} + F_n,$$

and the generalised Fibonacci sequence S_n is given by

$$S_0 = 0, \quad S_1 = 1, \quad S_{n+2} = kS_{n+1} + S_n,$$

where k is a positive integer. The case $k = 1$ gives the Fibonacci sequence, and $k = 2$ gives the Pell sequence.

The solution to the recurrence relation for S_n is

$$S_n = \frac{1}{\sqrt{k^2 + 4}}(\alpha^n - \beta^n),$$

where $\alpha = (k + \sqrt{4 + k^2})/2$ and $\beta = (k - \sqrt{4 + k^2})/2$, so that $\alpha\beta = -1$, $\alpha + \beta = k$ and $\alpha^2 + \beta^2 = k^2 + 2$. The identity

$$\left(\alpha^{2m} - \beta^{2m}\right)\left(\alpha^{2m+2} - \beta^{2m+2}\right)$$
$$= \left(\alpha^{2m+1} - \beta^{2m+1}\right)^2 - \alpha^{2m}\beta^{2m}(\alpha - \beta)^2$$

shows that $S_{2m} S_{2m+2} + 1 = S_{2m+1}^2$, and we deduce from the earlier result on triples of the form $(a, b, a + b + 2q)$ that

$$\left(S_{2m}, S_{2m+2}, S_{2m} + 2S_{2m+1} + S_{2m+2}\right) \tag{1}$$

is a Diophantine triple.

- *Show that* $(F_{2n}, F_{2n+2}, F_{2n+4})$ *is a Diophantine triple.*

25

Problem J: Discussion and Generalisations

First, we consider the case $n = 4$. The four points a_1, a_2, a_3 and a_4 form the midpoints of a quadrilateral with vertices v_1, v_2, v_3 and v_4 if and only if the v_j satisfy the simultaneous vector equations

$$\frac{1}{2}(v_1 + v_2) = a_1,$$

$$\frac{1}{2}(v_2 + v_3) = a_2,$$

$$\frac{1}{2}(v_3 + v_4) = a_3,$$

$$\frac{1}{2}(v_4 + v_1) = a_4.$$

If there is a solution then $a_1 + a_3 = a_2 + a_4$ or, equivalently, $a_1 - a_2 = a_4 - a_3$, which says (and you should prove this) that the a_j are the vertices of a parallelogram. If this is so, then given any b, there is a quadrilateral that is a solution with a vertex at b. We are now in a position to make a conjecture.

- *Conjecture: suppose that we are given distinct points $a_1, \ldots a_n$. If n is odd there is a unique n-gon with the a_j as the midpoints of the sides of the n-gon. If n is even, and if a certain linear combination of the a_j is zero, then infinitely many solutions exist. If this linear combination is not zero, then there is no solution.*

Obviously, we want to know more about the linear combination in the case when n is even. When $n = 2$ it is $a_1 - a_2 = 0$; when $n = 4$ it is $a_1 - a_2 + a_3 - a_4 = 0$.

- *What do you think it is in the general case?*

We shall now give a complete solution to the problem. This solution is algebraic, but is based on a geometric idea. It is valid in all dimensions, and it does not require the midpoints a_j, or the vertices v_j, to be coplanar. Given $a_1, \ldots, a_n$ we construct a map $f(x)$ as follows. Start with x and reflect x through the point a_1 to obtain $x_1 = 2a_1 - x$. Now reflect x_1 through the point a_2 to get

$$x_2 = 2a_2 - x_1 = 2(a_2 - a_1) + x.$$

Next, reflect x_2 through a_3 to get x_3, where

$$x_3 = 2a_3 - x_2 = 2(a_3 - a_2 + a_1) - x.$$

We continue this process until we have reflected through the point a_n, and the resulting point is $f(x)$. Thus

$$f(x) = \begin{cases} 2(a_n - a_{n-1} + \cdots + a_1) - x & \text{if } n \text{ is odd}; \\ 2(a_n - a_{n-1} + \cdots - a_1) + x & \text{if } n \text{ is even}. \end{cases}$$

The geometry of the construction implies that *there is a solution to our problem of midpoints that is a polygon with a vertex at x if and only if x is a fixed point of f* (that is, if and only this sequence of reflections ends at the starting point).

The following statements are now obviously true:

(1) if n is odd there is a unique solution to the problem, and one vertex of the solution is $a_n - a_{n-1} + \cdots + a_1$;

(2) if n is even there is a solution to our problem if and only if $a_n - a_{n-1} + \cdots - a_1 = 0$ or, equivalently,

$$a_1 + a_3 + \cdots + a_{n-1} = a_2 + a_4 + \cdots a_n.$$

In addition, if this condition is satisfied, then there is a solution with a vertex at any given point b.

With hindsight, a solution of this type is predictable. As the reflection through a point is a linear map (between vector spaces), the original question is equivalent to asking whether or not the linear map has a fixed point. *It is always useful to review a solution. Often, a solution looks much easier, or can be made much simpler, after we have struggled to solve a problem. We should always look again at a solution with a fresh mind.*

- *How can we generalise this problem further?*

There are several ways to generalise the problem. First, we note that the only mathematics that we have used is the theory of vector spaces so *the problem, and our solution, makes sense in any vector space (of any dimension)*. It holds, for example, in the (infinite dimensional) vector space of all real polynomials. Next, we can generalise from the case of a triangle in the Euclidean plane $\mathbb{R}^2$ to a tetrahedron in $\mathbb{R}^3$, but here there are two possible generalisations.

- *Given six points in $\mathbb{R}^3$, is there a tetrahedron for which these six points are the midpoints of the six edges of the tetrahedron?*

- *Given four points in $\mathbb{R}^3$, is there a tetrahedron in $\mathbb{R}^3$ such that the four given points are the centres of gravity of the four faces of the tetrahedron?*

We leave the first problem for the reader to consider, and we give a solution to the second problem. Suppose that the four given points are a_1, a_2, a_3, a_4. Then a tetrahedron exists with vertices v_j, and with the a_j as centres of gravity of the faces (labelled so that a_j is in the face opposite the vertex v_j) if and only if

$$v_1 + v_2 + v_3 = 3a_4,$$
$$v_2 + v_3 + v_4 = 3a_1,$$
$$v_3 + v_4 + v_1 = 3a_2,$$
$$v_4 + v_1 + v_2 = 3a_3.$$

The question now is *can we solve these simultaneous equations for the vectors v_j?* In fact, these equations have a unique solution, namely

$$v_j = (a_1 + a_2 + a_3 + a_4) - 3a_j, \quad j = 1, 2, 3, 4,$$

so there is a unique solution to this problem. If we want to be more ambitious we can speculate about cuboids and other solids in $\mathbb{R}^3$, and then move on to discuss similar problems in higher dimensions.

26

Problem K: Discussion and Generalisations

We recall that $X_n = t^{-1}\left(1 - \sqrt{1 - t^2}\right)$, where $t = 10^{-n}$, so we should consider the Taylor series for $\sqrt{1-x}$, and then put $x = t^2$. Since we want an accurate answer, and not an approximate answer, we should use a *finite* Taylor expansion.

Suppose that f is at least four times differentiable on the interval $(-1, 1)$, and that $0 \le x < \frac{1}{2}$. Then, for some y with $0 \le y \le x$, we have

$$f(x) = f(0) + f^{(1)}(0)x + f^{(2)}(0)x^2/2! + f^{(3)}(y)x^3/3!.$$

If $f(x) = \sqrt{1-x}$, we get

$$\sqrt{1-x} = 1 - \frac{x}{2} - \frac{x^2}{8} - \frac{x^3}{16(1-y)^{5/2}}.$$

As $0 \le y \le \frac{1}{2}$, we see that $16(1-y)^{5/2} \ge 1$, so that

$$\frac{x}{2} + \frac{x^2}{8} \le 1 - \sqrt{1-x} \le \frac{x}{2} + \frac{x^2}{8} + x^3.$$

If we now put $x = t^2 = 10^{-2n}$, we see that the conjecture is correct.

Index

stabiliser, 89
state
  odd, 46
  even, 46
switching lights, 22, 45

tetrahedron, 32
tower, 33, 75, 77

towers, 20
towers of positive integers, 20
triangle, 19, 71
triangle inequality, 29
triomino, 23, 51, 94

weighing chemicals, 23
weights, 23